85 Topics in Current Chemistry

Fortschritte der Chemischen Forschung

Instrumental Inorganic Chemistry

Springer-Verlag Berlin Heidelberg GmbH 1979

This series presents critical reviews of the present position and future trends in modern chemical research. It is addressed to all research and industrial chemists who wish to keep abreast of advances in their subject.

As a rule, contributions are specially commissioned. The editors and publishers will, however, always be pleased to receive suggestions and supplementary information. Papers are accepted for "Topics in Current Chemistry" in English.

ISBN 978-3-662-15822-7   ISBN 978-3-540-35247-1 (eBook)
DOI 10.1007/978-3-540-35247-1

Library of Congress Cataloging in Publication Data. Main entry under title: Instrumental inorganic chemistry. (Topics in current chemistry ; 85) Bibliography: p. Includes index. CONTENTS: Wolf, G. K. Chemical effects of ion bombardment. – Kiser, R. W. Doubly-charged negative ions in the gase phase. – Schwedt, G. Chromatography in inorganic trace analysis. 1. Chemistry, Analytic – Addresses, essays, lectures. 2. Chemistry, Inorganic – Addresses, essays, lectures. I. Wolf, Gerhard K., 1935 – Chemical effects of ion bombardment. 1979. II. Kiser, Robert Wayne, 1932 – Double charged negative ions in the gas phase. 1979. III. Schwedt, Georg, 1943 – Chromatography in inorganic trace analysis. 1979. IV. Series. QD 1.F 58 vol. 85 [QD 75.2] 540'.8s546 79-14 180

© by Springer-Verlag Berlin Heidelberg 1979
Originally published by Springer-Verlag Berlin Heidelberg New York 1979.
Softcover reprint of the hardcover 1st edition 1979

2152/3140–543210

Contents

Chemical Effects of Ion Bombardment

Gerhard K. Wolf

Physikalisch-Chemisches Institut, Universität Heidelberg, Germany

Table of Contents

G. K. Wolf

1 Introduction and Survey

The techniques of ion bombardment and ion implantation are receiving increasing interest in different fields of science. Their application for the solution of problems in basic science and their potential utilization in applied science are numerous. Some years ago the bombardment of semiconductors was the only widely used application. While this is still the field of the most practical importance the implantation of materials in order to improve their resistance against mechanical and chemical attack is receiving more and more attention.

An energetic particle hitting a surface or interacting with a gas may
— merely cause damage,
— itself act as an important quantity for the process,
— end up as a reaction partner during the formation of a new alloy or compound.

Table 1 represents an attempt to classify the studies using ion beams according to their field of science and according to the main effects caused by the ions. The arrangement is somewhat arbitrary. But because of the great overlap between the different kinds of studies, this cannot be avoided.

Therefore, ion implantation is a real interdisciplinary field in which engineers as well as physicists and chemists work. The chemists were in a minority until recently and the chemical aspects of ion implantation received much less attention than the physical and technological ones. Nevertheless, they are just as interesting and of practical importance for its future development. For these reasons it seems justified to present the most important works with chemical significance done during the last decade by chemists, physicists and engineers.

Unfortunately, it is very difficult to demarcate the area of ion bombardment from the surrounding area and to distinguish the more chemical effects from the other effects induced by bombardment. Therefore one has to try to find a definition of "chemical effects of ion bombardment" in the sense of the present article. It includes accordingly all studies concerned with:

a) Chemical effects of the radiation or, in other words, changes in the chemical structure and composition of the targets induced by the bombardment. The relevant chapters are "Radiation Damage" and "Surface Chemistry of Materials Irradiated with Ions".
b) The chemical state and the chemical reactions of the bombarding ions themselves. Examples are found mainly in the chapters "Solid State Chemistry of Implanted Atoms" and in "Surface Chemistry of Materials Irradiated with Ions".
c) The use of implanted ions as a probe for the state of a system. The chapters "Solid State Chemistry of Implanted Atoms" and "Radiation Chemistry" contain a few examples of interest for chemists.
d) The applications of ion-irradiated materials in chemistry independent of the irradiation effects being of chemical or physical nature. These studies are mentioned in the chapter "Surface Chemistry of Materials Irradiated with Ions".

Semiconductor and thin film technology is an area too large to be included in this survey, apart from a few comments in Sect. 5. This is also true for "Implantation Metallurgy", which has been covered by a number of recent review papers[1, 2].

Table 1. The application of ion beams in different areas of science

Effects caused by the bombardment	Solid state physics	Electrical engineering	Material science	Solid state chemistry	Surface chemistry and physics	Atomic physics physical chemistry
Radiation damage	Superconductivity of bombarded metals	Variations of thin film properties	Simulation of radiation damage e.g. in nuclear power plants; Wear, friction and lubrication of materials	Radiation Chemistry	Ion sputtering; Ion reflection; Reactivity of ion-bombarded surfaces	Radiation decomposition of gases; Ionization phenomena; Charge exchange studies
Insertion of impurities	Superconductivity of bombarded metals; Doping of optical materials	Variations of thin film properties; Doping of semiconductors	Wear, friction and lubrication of materials; Formation of non-equilibrium alloys	Chemical state of implanted atoms	Reactivity of ion-bombarded surfaces; Corrosion of materials; Electrochemical properties of materials	
Reactions			Formation of non-equilibrium alloys	Syntheses of non-conventional compounds; Simulation of "Hot Atom" reactions; Simulation of high-energy reactions in space		Simulation of "Hot Atom" reactions; Ion-molecule reactions

The radiation effects and ion-molecule reactions in gases are other topics which, strictly speaking, belong to the subjects of this article. But being much too voluminous the subject would by far exceed the extent of this paper. Fortunately, there are several relevant books available[3, 4].

Since there exist quite a number of other methods for surface modification and for introducing foreign atoms into host lattices, one should discuss briefly the advantages of ion implantation:

a) It is possible to dope every host material with every element. There is no need to consider the question of solubility or other thermodynamic quantities.

b) Ion implantation is the only method allowing the doping of a host material at any temperature one likes. Studies of the recovery behavior of materials in the low-temperature region and of the low-temperature chemistry of foreign atoms in hosts are thus possible without prior treatment of the system.

c) One may study the behavior and the reactions of single atoms because the number of ions interacting with the solids or gases may be limited to avoid mutual influence.

d) Ion bombardment of solids leads to variations of very thin surface layers of a few Å to a few 100 Å, while the bulk of the material remains unchanged.

e) The density of defects in the implanted region of solids is very high.

The last two arguments are controversial. In certain cases the impossibility of doping the bulk of a material or the creation of defects through the implantation could be disadvantageous. This statement indicates the limitations of the method as well as the fact that it is an experimentally pretentious technique.

The following three chapters provide an outline of the fundamentals of the interactions of heavy ions with matter and a description of the most important experimental methods for the production and acceleration of ions as well as for the analysis of the products of the bombardment.

In Chaps. 5 and 6 the experimental work with chemical relevance to radiation effects as well as the chemical "fate" of the bombarding particles and the applications of the irradiated material are covered.

The last chapter proposes some future trends in this field.

2 Fundamentals of the Interaction of Heavy Ions with Matter

2.1 Energy Loss

An energetic ion passing through matter loses energy through interaction with the surrounding atoms. The mechanism of this energy loss was the subject of many years of investigation, starting in 1913 with Bohr[5] and most completely developed in 1963 by Lindhard[6]. The treatment by different authors contains many assumptions and correction factors, but the agreement within the experiments is rather good in general, taking into account a number of additional processes not included in the theory.

The following summarizes the theory of Bohr, Bethe and Lindhard et al. The major processes of energy loss are:

- Excitation of the electrons of the surrounding atoms by the energetic ion. This process is called electronic energy loss or electronic stopping.
- Collisions of the ions with the atoms, called nuclear energy loss or nuclear stopping.
- Charge exchange between the ions and the atoms.

The total energy loss is the sum of the three fractions, and may be written

$$\left(\frac{dE}{dx}\right)_{total} \doteq \left(\frac{dE}{dx}\right)_{electronic} + \left(\frac{dE}{dx}\right)_{charge\ exchange} + \left(\frac{dE}{dx}\right)_{nuclear}$$

Since charge exchange represents only a very small fraction of the total energy loss, it will be neglected in the following considerations.

Electronic energy loss is the prevailing fraction at high energies or low mass numbers; nuclear energy loss dominates at low energies and high mass numbers. Thus He ions of 100 keV lose their energy nearly exclusively by electronic excitation, 10 keV Xe ions mainly by elastic collisions.

At even higher energy another mechanism starts to play a role, namely Rutherford scattering at the atomic nuclei. But for the energy region we deal with it has not to be taken into account.

2.1.1 Electronic Stopping

In the theory of Bohr[7] the energy loss of an ion interacting with an atom is

$$\left(\frac{dE}{dx}\right)_{electronic} = \frac{4\,\pi\,Z_1^2\,e^4}{mv^2}\,B$$

This is only valid for a fully ionized atom moving with a higher velocity than the K-shell electrons. B is an interaction parameter for which Bethe[8,9] gave the expression

$$B = Z_2 \ln\left(\frac{2\,mv^2}{I}\right)$$

Z_1, Z_2 = nuclear charge of the ion and atom
m = electronic mass
v = projectile velocity
I = average excitation energy of the atom

For lower velocities one has to apply inner shell corrections as well as for the capture of electrons by the primary ion[10 – 12].

At low velocities $(v_1 \ll Z_1 \cdot v_0; v_0 = Z_1 e^{2/\hbar})$ the electronic stopping power is proportional to v_1.

For this region Lindhard and Scharff[13] gave the formula

$$\left(\frac{dE}{dx}\right)_{\text{electronic}} = \xi \, \frac{8 \, \pi \, e^2 \, N \, a_0 Z_1 Z_2 \, v}{e(Z_1^{1/3} + Z_2^{2/3})^{3/2} \, v_0}$$

$\xi \quad \sim \, Z^{1/6}$
$a_0 \quad = \text{first Bohr orbit}$
$N \quad = \text{number of atoms per unit volume}$
$v_0 \quad = \text{Bohr velocity}$

2.1.2 Nuclear Stopping

Nuclear stopping is only important at low velocities ($v_1 < Z_1 \cdot v_0$) where collisions between the projectile and the target atom as a whole take place. They can be treated using classical mechanics if the energy is above a certain value, a condition being fulfilled nearly in all cases of interest for the present paper. The most difficult task in this treatment is the proper choice of an interatomic potential between two colliding atoms. Bohr[14] as well as other authors discussed this question. There have been several attempts using Born-Mayer-, Coulomb-, Nielson-, Brinkman-, and Fiersov-potentials[15].

Lindhard et al. used a Thomas-Fermi potential and calculated the differential scattering cross section for multiple collisions[6] as

$$\sigma = \frac{\pi \cdot a^2}{2 \, t^{3/2}} \, f(t^{1/2})$$

where f is a numerically evaluated function for which Winterbon et al. give an analytical approximation[16], and t a dimensionless variable and a a measure for the size, connected to the first Bohr orbit a_0:

$$a = \frac{a_0 \cdot 0.8853}{(Z_1^{2/3} + Z_2^{2/3})^{1/2}}$$

2.1.3 Energy Loss Expressed in Universal Constants

The most interesting feature of the work of Lindhard et al.[6] is the possibility of expressing the electronic and the nuclear stopping power in terms of universal constants, a form independent of the mass and charge of the individual atoms. A "universal" energy ϵ and a universal range ρ are defined, and the energy loss is expressed in terms

of $\left(\dfrac{d\epsilon}{d\rho}\right)$ instead of $\left(\dfrac{dE}{dx}\right)$.

The electronic energy loss takes the form

$$\left(\frac{d\epsilon}{d\rho}\right)_{electronic} = k\,\epsilon^{1/2}$$

with $k \approx \dfrac{0.079\,Z_1^{1/6}Z_1^{1/2}Z_2^{1/2}\,(M_1 + M_2)^{3/2}}{(Z_1^{2/3} + Z_2^{2/3})^{3/4}\,M_1^{3/2}\,M_2^{1/2}}$

The nuclear energy loss is related to the stopping cross section σ by

$$\left(\frac{d\epsilon}{d\rho}\right)_{nuclear} = \sigma\,\frac{(M_1 + M_2)}{4\,\pi\,e^2\,Z_1Z_2M_1}\ , \text{ where}$$

the universal energy is calculated by

$$\epsilon = E\,\frac{a\,M_2}{Z_1 Z_2 e^2(M_1 + M_2)}$$

and the universal range is calculated by

$$\rho = RNM_2\,\frac{4\,\pi\,a^2\,M_1}{(M_1 + M_2)^2}$$

M_1, M_2 represent the masses of the projectile or target respectively,
R is the normal range of the projectile in matter and
N is the number of "stopping atoms" per unit volume.

Figure 1 shows the nuclear and electronic specific energy loss.

From the curves it is possible to calculate the "normal" stopping using the above equations. Note that the curve for the nuclear stopping is really universal, whilst the curve for the electronic stopping has a different slope (k) for every combination projectile-target. The figure shows very clearly that nuclear stopping is the dominant mechanism at low energies and electronic stopping at high energies.

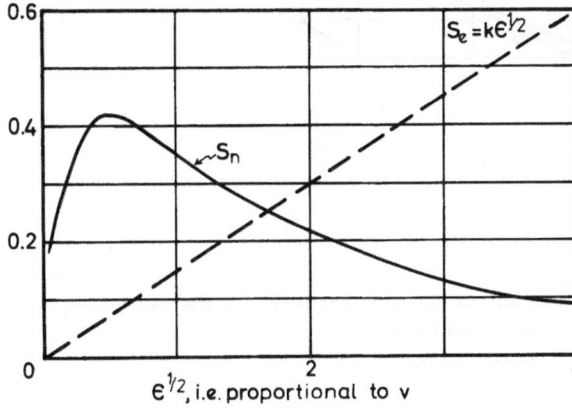

$\epsilon^{1/2}$, i.e. proportional to v

Fig. 1. Nuclear and electronic specific energy loss S as a function of energy in dimensionless units

2.2 Ion Ranges and Profiles

2.2.1 Ranges

The treatment covered in the preceding chapter allows the prediction of ranges of implanted atoms. The range is the total path of the ion or the integral over all $\dfrac{dE}{dx}$.

If one uses the expression range, one has to define the kind of range meant. Figure 2 tries to clarify the different types. One might be interested in the total range R_L or, more often, in the projected range R_p which is a measure of the distance of the ion from the surface and has the advantage of being easily accessible to measurements.

An approximate relation for the region of nuclear stopping is given by Lindhard[13]

$$\frac{R_L}{R_p} = 1 + \frac{M_2}{3 M_1}$$

The electronic stopping does not lead to major deflection and does not contribute very much to the lateral spread $R_\perp$. Another important quantity is the range straggling ΔR or ΔR_p. It contains information about the actual distribution of the implanted ions in a solid. They also have been calculated by various authors, e. g. Schiott[18].

In practice, one normally uses tables or graphs presenting ranges, range straggling, and lateral spread as a function of projectile-target combination and energy. In general, the energy is given in keV and the range or projected range in $\mu g\ cm^{-2}$, which allows an easy comparison of different materials. Some range tables are presented in Table 2.

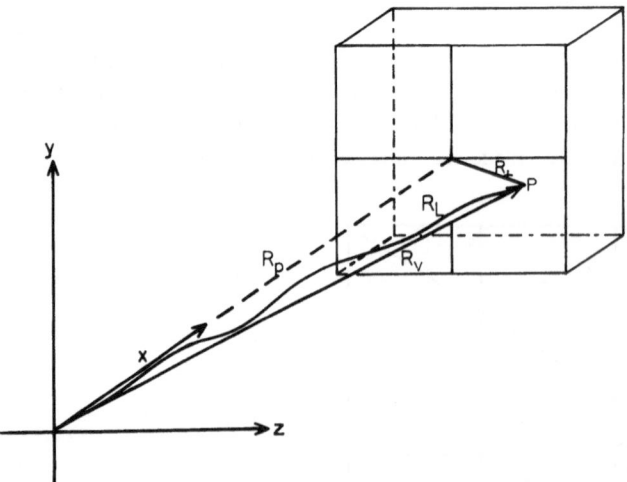

Fig. 2. The different types of ranges for an ion starting from x=y=z=0[17] in the x direction

R_L = total range, $\qquad$ R_p = projected range,
R_v = vector range, $\qquad$ $R_\perp$ = lateral range,
$R_v^2 = R_p^2 + R_\perp^2$

Table 2. Range-energy tables

Author	Quantities listed	Targets	Ref.
G. Dearnaley et al.	$R_p + \Delta R_p$	$Z_1, Z_2 = 5, 10, 15 \ldots$	19)
J. Gibbons et al.	$R_p, \Delta R_p$ $R_L, \Delta R_L$ $\dfrac{dE}{dR}_{nucl.}, \dfrac{dE}{dR}_{el.}$	Selected elements and compounds	20)
R. G. Wilson, G. R. Brewer	$R_p + \Delta R_p$	Selected elements and compounds	21)

Another possibility is the use of graphs where the reduced range is plotted versus the reduced energy ϵ, as derived from the theory of Lindhard. They have the advantage of containing all possible information on one page, the disadvantage that one has to calculate the "normal" units from the equations given in the preceding section. In most cases the values given in the tables are only for elemental targets or semiconductor compounds. For other compounds one may estimate the ranges assuming a fractional contribution from the different elements.

$$\frac{1}{R_p} = \sum_i \frac{x_{1,2\cdots i}}{R_{p1,2\cdots i}}$$

$R_{p1, 2 \ldots}$ = Range of the elements 1, 2 ...
$x_{1, 2 \ldots}$ = Fraction of the component in question at the total mass

The accuracy of this estimate is about ± 10 %[22].

2.2.2. Range Profiles

Range or concentration profiles contain information about the distribution of the bombarding ions in the solid target. Since the mapping of the distribution in three

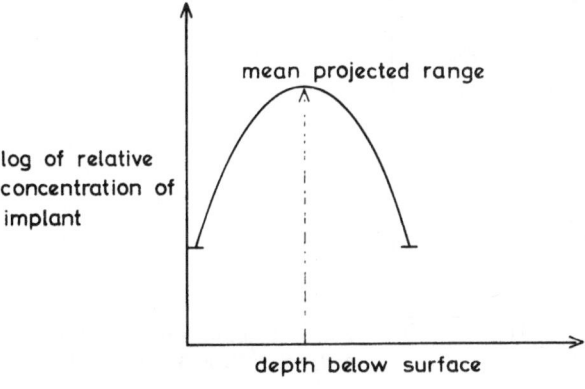

Fig. 3. Theoretical concentration profile of implanted atoms in a solid target

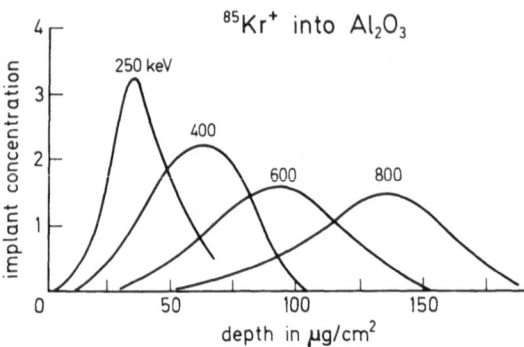

Fig. 4. Range distribution of ^{85}Kr ions implanted at different energies in Al_2O_3[23, 24]

dimensions is quite intricate, one normally presents the intensity distribution in the direction normal to the surface. This gives a picture of the actual depth distribution of the ions in the target. The theoretical range profiles are very similar to a Gaussian distribution around the mean projected range (Fig. 3).

The energy of the ions is responsible for the penetration depth and the depth distribution.

Figure 4 shows the experimental distribution of ^{85}Kr ions implanted into amorphous Al_2O_3 at different energies.

Such distribution data can also be presented as integral curves. Here the concentration remaining beyond a certain value at any penetration depth is plotted. This kind of description makes it very easy to deduct the variations between two depth values or the concentration at high depth values. Figure 5 shows the same distributions as Fig. 4 in the integral form.

The range profiles measured experimentally very often show considerable deviations from the Gaussian form. The reasons for these deviations are:

a) In single crystals or polycrystalline material an alignment of the ion beam with a crystallographic axis leads to an enhanced penetration of the ions due to the channeling effect (Sect. 2.4). Figure 6 is a presentation of results using such a system.

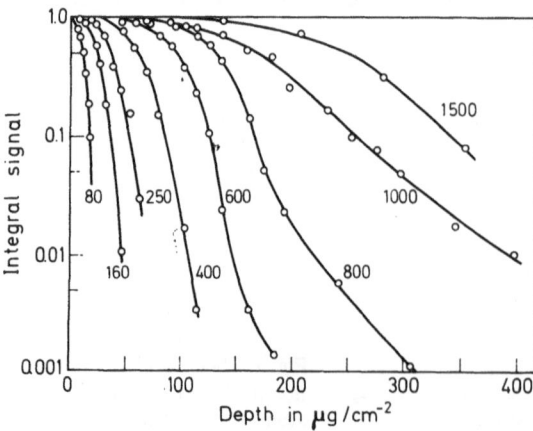

Fig. 5. Range distribution (integral) for ^{85}Kr in Al_2O_3[23, 24]

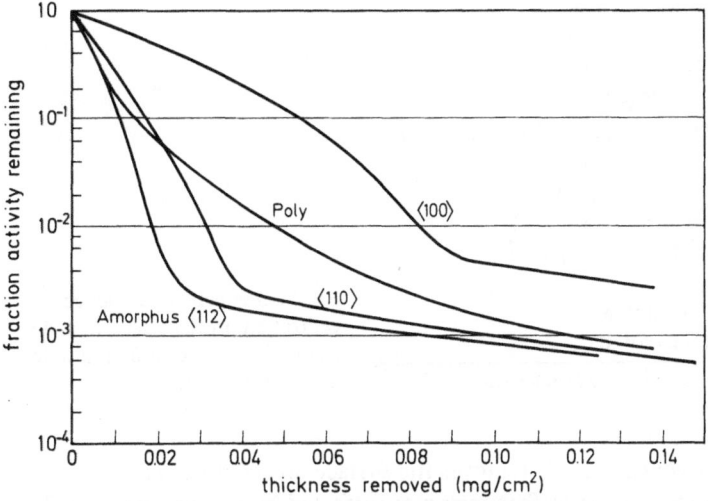

Fig. 6. Range profiles of ^{85}Kr in amorphous, polycrystalline and single crystal tungsten[25]

An increase in dose slowly destroys the ordered crystal structure, leading to a distribution similar to the amorphous case. The channeling effect in polycrystalline material is much smaller compared with single crystals.

b) At elevated temperature the diffusion of the implanted atoms very often leads to a broadening of the profile. In addition diffusion processes play a role at low temperatures during the irradiation, leading to "radiation enhanced diffusion" (Sect. 2.6). Fig. 7 illustrates the effect.

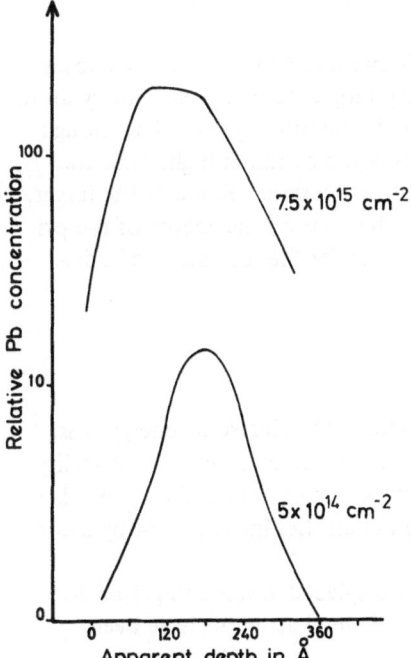

Fig. 7. Broadening of the concentration profile due to radiation-enhanced diffusion (after[26])

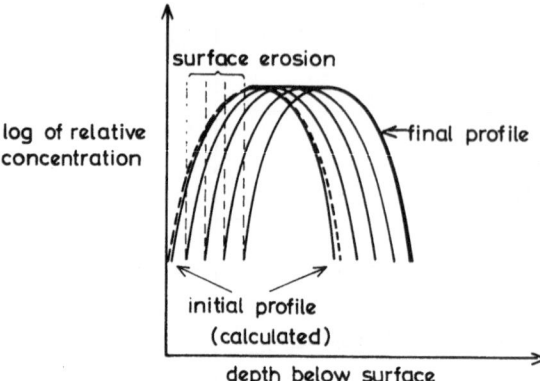

Fig. 8. Originally Gaussian concentration profile after a high-dose irradiation. The erosion is due to the sputtering effect

c) The sputtering effect (Sect. 2.5) removes the surface layers of the target. Naturally the implanted ions of the surface region are affected by this process, too. This leads to a modification of the profile as shown in Fig. 8.

Without going into further details, it should be mentioned that a number of techniques exist for shaping the concentration profile or achieving a more uniform profile:
- subsequent bombardments with different energies,
- controlled diffusion of the implanted species,
- erosion of parts of the profile due to sputtering,
- bombardment of the target at different angles.

2.3 Energy Distribution

Neglecting the electronic energy loss, we may assume that a primary particle loses a fraction of its energy in colliding with a secondary target atom. The secondary atom may have enough energy to displace a third atom. If the primary atom has enough energy, the number of displaced secondary atoms could be rather high. Thus the energy is dissipated by a collision cascade containing a certain volume of the target. The number of displacements (N) in the cascades depends on the energy of the primary particle (E_{pr}) and the minimum energy required for the formation of a displacement (E_d).

$$N = \frac{E_{pr}}{2\,E_d} \quad (E_{pr} \gg E_d)$$

This expression gives only a rough estimate. It neglects the electronic energy loss, the form of the atomic interaction and replacement collisions. Therefore, in reality, the above formula will give a considerable overestimation of the number of displacements. It is possible to take some of the neglected quantities into account by a correction factor.

One can calculate the complete profile of the displaced atoms and get in that way a spatial distribution of the deposited energy. Several attempts have been made[27, 27a].

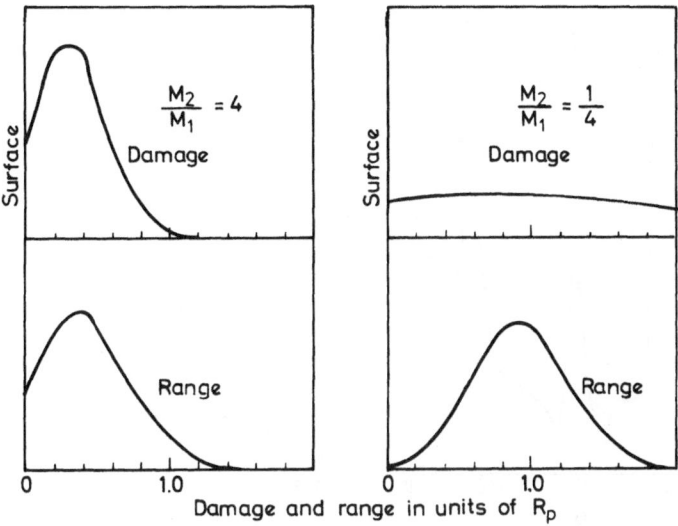

Fig. 9. Computed range and damage profiles for different mass ratios (after[28])

There are cases where the range of the displaced atoms exceeds the range of the implanted ions and vice versa, depending on the ratio of the mass M_1 of the ion to the mass M_2 of the displaced atom. Fig. 9 presents the range of the ions and the corresponding energy distribution for

$$\frac{M_1}{M_2} = \frac{1}{4} \quad \text{and} \quad \frac{M_1}{M_2} = 4$$

Looking at this presentation, one should keep in mind that the experimental energy distributions may deviate a good deal from the calculations.

2.4 The Channeling Effect

The enhancement of ion or damage penetration in ordered materials along certain crystallographic directions has already been mentioned in the preceding section. There exist two main reasons for this effect:
- energy may be transported along close-packed rows of a crystal by focusing collision sequences far away from the initial event. This so-called "focusons" very often accompany a mass transport because of the last atom of the sequence being displaced from its initial position. This results in the damage proceeding in a favored direction.
- The primary implanted atoms may propagate much further into the crystal by following the direction of a "channel". The host atoms, forming the "walls" of the channel, give rise, due to their potential distribution, to a focusing effect along the centre of the channel. The consequence is that channeled atoms are found up to 10 times further inside a crystal than random ones[29].

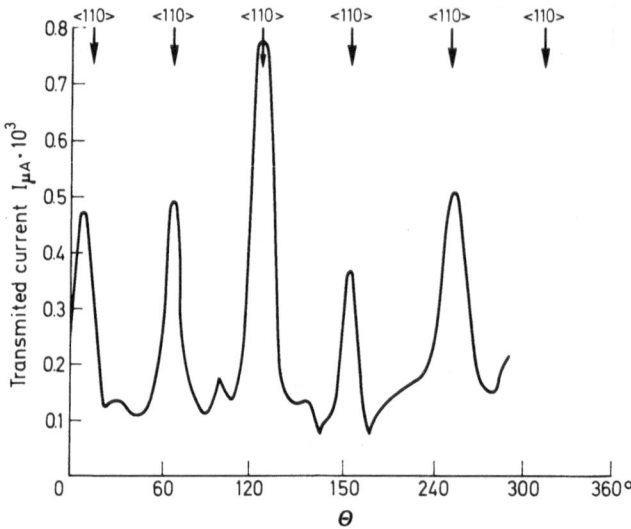

Fig. 10. Variation of the transmitted proton current with the rotation of a single crystal about the (111) axis

The channeling effect is illustrated experimentally by letting protons penetrate through a thin aligned gold crystal. The proton current, a measure of the transmission probability, is a strong function of the orientation of the crystal. In Fig. 10 the transmitted ion current is plotted as a function of the rotation of the crystal around the (111) axes[30].

An effect related to channeling is called blocking. If a displaced atom or an impurity in an interstitial position "blocks" a channel, the preferential penetration of atoms through this channel is hindered. Therefore, the presence of damage or impurities has some influence on the channeling effect, a fact being used in backscattering experiments for the location of displaced or foreign atoms (Sect. 4.2.1). Also the regular atoms in an atomic row of the lattice may block the penetration of ions if the crystal is misaligned. While the maxima of Fig. 9 correspond to favored directions of channeling, the minima show favored directions of blocking. One may also record the intensity of the proton or ion beam transmitting a thin crystal in a two dimensional representation. The screen or photograph will show a regular pattern of dark and light spots or lines corresponding to the direction of favored channeling and blocking of the scattered particles. Such a pattern is useful for getting information on the structure of a crystal.

2.5 Sputtering

In Chap. 2.3 the dissipation of the energy of the primary events by a collision cascade was mentioned. If a primary collision takes place near the surface, a part of this cascade may be directed to the surface. This leads eventually to the ejection of surface atoms from the solid as shown in Fig. 11.

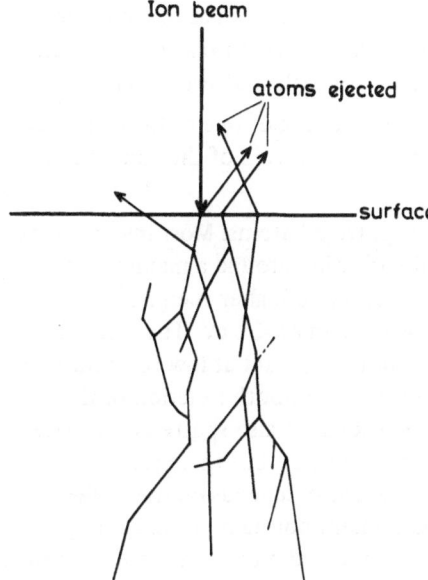

Ion beam

atoms ejected

surface

Fig. 11. A collision cascade leading to the ejection of surface atoms

In general, one characterises the sputtering process in terms of the sputtering yield. The sputtering yield is the number of atoms ejected from a surface due to the incidence of one bombarding ion. The sputtering yield is dependent on several variables. The most important ones are the energy and the mass of the bombarding ions. The yield increases with the mass and with the energy until a maximum is reached, typically in the region of $10^4 - 10^5$ eV.

In addition to these factors, the elemental composition of the target plays a role. Tantalum or similar metals as targets have sputtering yields of $\leqslant 1$ atom/ion, whereas metals like Ag, Au and Cd may have values from $10 - 100$ atoms/ion.

Experimentally measured sputtering yields, like the ones reported by Almen and Bruce[31], also show a Z dependence (Z of the ion) in addition to the mass dependence and have yield peaks around bombarding elements like Cd and Hg. These are no "real" peaks but, due to the implantation of the bombarding ions, the surface is altered under formation of phases or alloys. Thus after a certain period of bombardment the alloy is sputtered instead of the original target. Taking into account such experimental difficulties, one finds a rather good agreement between the experiment and the sputtering theory. Among the various attempts made to describe this phenomenon the Sigmund theory, in particular, should be mentioned[32].

The sputtering effect has a number of practical consequences for ion bombardment experiments.

a) As already mentioned, sputtering causes surface etching and, therefore, changes the range profile at high implantation doses.

b) Sputtering limits the maximum amount of a certain ion which can be implanted into a surface. After an elongated bombardment the number of atoms already implanted and sputtered again from the surface is just as high as the number of the freshly incoming atoms. The equilibrium dose is $\geqslant 10^{18}$ ions/cm^2 for materials with low sputtering yield and around 10^{16} for high sputtering yield.

c) Alloys or compounds may suffer preferential sputtering of one of the elements. Therefore, during the bombarding process the composition of the compound may change.

Not much is known about the energy of the sputtered atoms. Most investigators come to the conclusion that the atoms show a distribution around a maximum intensity of $1 - 10$ eV which extends with a E^{-2} dependence to higher energies[33]. At elevated temperatures the maximum shifts to lower energies < 1 eV. The idea of thermal spike evaporation, a mechanism which is not important at low temperatures, developed from this finding. The energy distribution is a smooth function of the energy and mass of the primary ions. Besides atoms ions are also sputtered. In general, their yield is orders of magnitude lower than the neutrals, but there are certain exceptions like the alkali halides. The sputtering yield also depends on the angle of incidence of the beam to the target. The yield maximum normally occurs at angles between 50° and 80°, at higher angles the beam is reflected more and more from the surface and the yield decreases.

There exist a number of investigations dealing with surface topography as a result of sputtering which we cannot treat in detail. Generally speaking, one would expect a smoothing of the surface. In practice this is only true for planar isotropic material. In a number of cases a new structure is produced:
— an impurity with a lower sputtering yield may protect the underlying material,
— the initial surface topography may shadow certain regions of the surface, especially in the case of a sloping incidence of the beam,
— there will be a flux enhancement at edges and steps of the material,
— in craters the atoms already sputtered are deposited again,
— in crystalline solids grains with different crystallographic orientation are sputtered at different rates. Grain boundaries and dislocations may also play a similar role.

Summarizing, one may say that sputtering has important consequences for the bombarded material and the surrounding area. It is widely applied for polishing and surface cleaning as well as for coating and thin film production. Some of its chemical aspects will be treated in Chap. 5.1.

2.6 Defects and Radiation Damage

The most important action of energetic ions upon materials is the radiation damage leading to the production of fragments and defects. Since the former are only important in the gas phase and in covalent compounds, we shall cover them in Chap. 5.1. In this section only the defects produced in solids will be considered.

The formation of defects may be either the desired effect of a bombardment or only the undesirable by-product of an implantation. In both cases some knowledge about the structure and behavior of defects is advantageous.

2.6.1 The Creation of Point Defects at Low Temperature

We mentioned above the collision cascade producing displaced atoms in a solid target. If we consider a single collision event in a crystalline solid, we can see that this displacement of atoms leads to the preferential formation of point defects. The most important types of point defects are:
a single vacancy (one atom or ion is missing),
a single interstitial (an additional atom),
a vacancy-interstitial pair (Frenkel pair).
In addition to these defects di-vacancies, agglomerations of interstitials and extended clusters of point defects are well known.

If one calculates the energy necessary for the formation of point defects, one ends up with typical values of $0.5 - 5$ eV[34]. Looking at ion implantation in practice, the situation is quite different. A dynamic situation exists because the bombarded lattice region is not in thermal equilibrium when the displacement occurs. Thus the threshold energy — the minimum energy required for the formation of a Frenkel pair — is above 10 eV. It is possible to determine this energy by irradiating the target with electrons (which only produce point defects) of increasing energy. At the threshold one starts to observe the formation of defects. Since the threshold is not a sharp step function, the average displacement energy for the formation of a Frenkel pair is higher. It depends on the mass of the atoms as well as on the binding energies and the crystallographic orientation. Typical values are 20 eV for Cu, 30 eV for Ge and 40 eV for Au. Normally a heavy ion transferring its energy to a lattice will cause several displacements. The most simple formula to estimate their number is

$$\bar{N}_{(E_0)} = \frac{E_0 - I}{2\,E_d} \quad (E_0 \geqslant E_d)$$

$\bar{N}$ = average number of displacements
E_0 = energy of the heavy ion
E_d = average energy required to displace a lattice atom
I = contains other energy loss mechanisms apart from displacement

This formula contains the assumption that energy transfer below the threshold for the displacement of one atom and between the threshold energy for the displacement of one atom and the threshold for the displacement of two atoms takes place via collisions which do not lead to the formation of point defects.

The expression "displacements per atom" (dpa) is a quantity very often used for comparing the damage caused by ion bombardment in different targets by different types of bombarding particles. In most cases the displacements per atom are normalized to the total number of target atoms or atoms in a certain region of the target.

Empirically the above formula considerably overestimates the average number of displacements for two reasons:
— a certain number of the Frenkel pairs formed recombine immediately and are therefore not accessible for experimental investigations. One may define a certain recombination volume which limits the minimum distance necessary for the formation of irreversible defects.

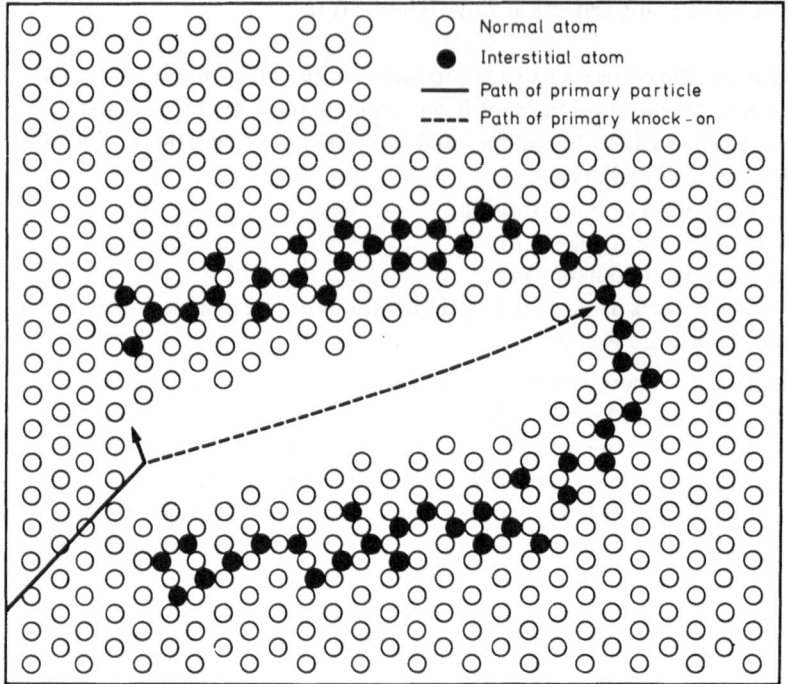

Fig. 12. Displacement spike arising from an energetic ion in a solid[35]

— focusing collisions along close-packed atomic rows and channeling dissipate the energy more than normal collisions. Fewer displacements per unit energy are the result. The consequence of both phenomena is that stable vacancies and interstitials have a separation distance of at least several atomic intervals. One thinks of a vacancy-rich region surrounded by a shell of material rich in interstitials. Such a displacement spike caused by an energetic ion is shown in Fig. 12[35, 36].

Another theoretical approach is the so-called "thermal spike" concept. The spike originates from the local heating of certain lattice regions due to the bombardment. A metal like copper requires only a fraction of eV/atom to raise its temperature above the melting point. The "melting" of a relatively large region of the target is thus possible. In practice the experimental results show that uniform melts certainly do not occur. Probably the short duration of $\approx 10^{-12}$ sec of the "high temperature" state does not allow a statistical treatment as required by the definition of temperature.

The spontaneous recombination of close pairs of point defects has another important consequence. As soon as the damaged regions caused by two bombarding ions overlap, recombination reactions take place. This means that the damage increase with increasing dose approaches a saturation value limiting the maximum obtainable stable defect concentration.

The situation becomes even more complicated going from elements, especially metals, to compounds. Here in addition to the effects arising from collisions, the excitation of electrons leading to ionization or excited states have to be taken into consideration. All these processes produce either defects or give rise to the breaking of bonds with subsequent rearrangement of the surroundings of the atom in question. In ionic crystals, for example, a number of additional point defects are known in comparison with metals. In alkali halides other than the types already mentioned, one observes

F-centres = an anion vacancy containing an electron

F'-centres = an anion vacancy containing 2 electrons

F_2-centres = two anion vacancies containing 2 electrons

H-centres = a Cl_2 molecule at an anion site or in other words, a·Cl^0 interstitial bound to a Cl^- ion.

Other point defects in ionic crystals are the

V-centres = a cation vacancy bound to one or more holes corresponding to the charge of the missing cation

Impurity centres = F, H or V centres trapped at an impurity cation or anion.

2.6.2 Secondary Processes at Higher Temperatures

The initial damage configuration of ion bombarded materials is only preserved at liquid He-temperature. Measurements at higher temperatures show the damage configuration being influenced mainly by annealing and diffusion.

Since the majority of all experiments are carried out at room temperature (RT), one does not know very much about the initial configuration and the recovery behavior in the low temperature region. In metals the recovery process is initially due to the mobility of the primary interstitials. In the region up to 100° K they recom-

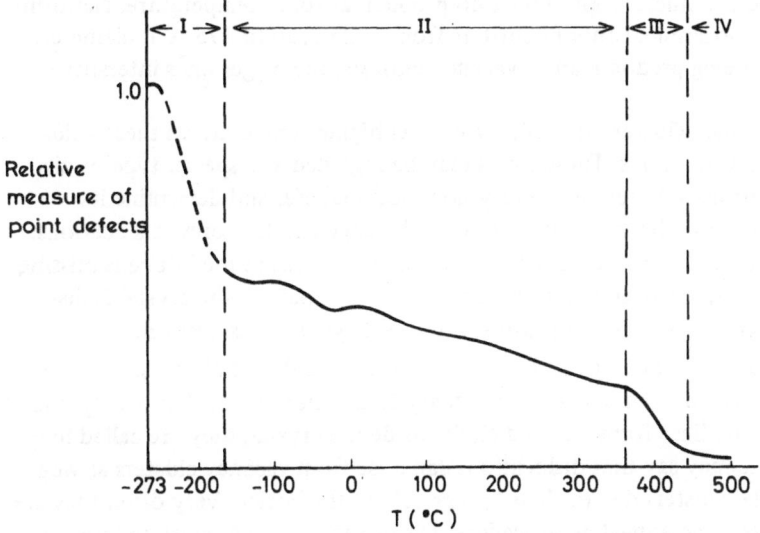

Fig. 13. Recovery curve for W irradiated with fast neutrons (after[37])

bine with close vacancies. In a second stage they diffuse further and recombine with more distant vacancies. Another process occurring in the low energy region is the trapping of interstitials leading to the formation of dumb-bell configurations being again mobile at higher temperature or the trapping at impurity sites. Finally near room temperature the vacancies also start to be mobile.

Fig. 13 shows the recovery of neutron-irradiated W as deduced from resistivity measurements. Irradiations with fast neutrons are within certain limits comparable to ion bombardments. One observes clearly the different stages I to III which one finds for most other metals shifted to either lower or higher temperatures.

The interpretation of these stages is still far from being clear. Most authors regard stage I as being due to recombination of close pairs and trapping of interstitials, stage II as being due to detrapping and clustering of interstitials, and stage III as being due to vacancy migration and trapping[38, 39]. The recovery of compounds is much more complicated and much less investigated. As compared to elements at least ionic crystals show some similarities with metals. Here also the bulk of the point defects seems to anneal at rather low temperatures[40, 41]. A consequence of the prompt radiation annealing and thermal annealing of interstitials is the "diffusion" of interstitials into the bulk of material. Thus a certain percentage of the point defects, as well as of the implanted atoms, come to rest far beyond the region of the normal distribution in an area of the lattice which has been only minimally damaged. This tail of the distribution may in certain cases contain up to 10% of the total bombarding particles.

2.6.3 Extended Defects and Precipitations

We have already mentioned clustering occurring during the recovery of irradiated material. Experimentally one finds interstitial and vacancy clusters, the size of which increases with temperature. While it is not possible to detect any visible clusters at $\sim 100°$K by electron microscopy, this is no problem at room temperature. Neutron-irradiated Cu, for example, contains clusters from ~ 25 Å up to 175 Å in diameter, the smaller ones being predominantly vacancy clusters, the bigger ones interstitial types[43].

The large vacancy clusters are called voids. At higher temperatures these voids may collapse and form loops. These loops may be regarded as a special type of dislocation. Dislocations are present in every non-ideal material and determine its mechanical properties. The two main types are the edge and the screw dislocations. Defects are called edge dislocations when one plane of atoms in the lattice is missing or supernumerary; screw dislocations are formed when a part of the crystal is displaced by an atomic layer. Fig. 14 illustrates the two types of dislocation.

In an ideal situation dislocation lines would penetrate the whole crystal. In reality they mostly extend from one grain boundary to another one or they are "pinned" by impurities. If the lines form a closed circle inside the crystal, they are called loops. Summarizing, one may say that dislocations can arise from vacancy clusters as well as from interstitial clusters due to their "pressure" on the lattice. Very often they are the final products of an annealing procedure. Dislocations already existing interact with point defects and impurities acting as traps or sinks.

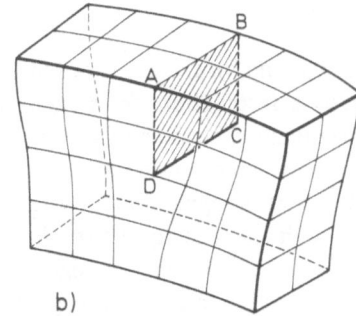

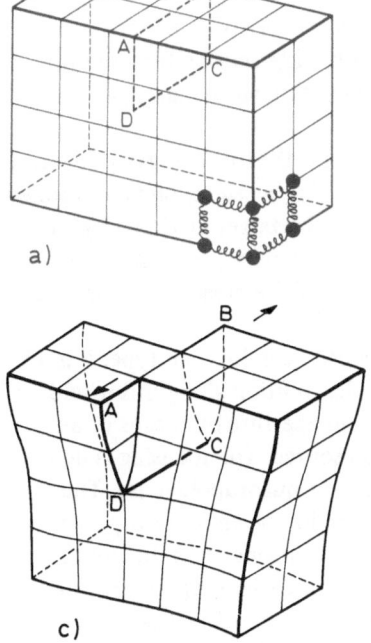

Fig. 14. Model of an edge (**b**) and screw (**c**) dislocation in a simple cubic lattice[44]

Up to now we have looked only at the effects of the bombardment and we have neglected the fate of the implanted atom itself. These atoms may be regarded as impurity atoms. During the recovery of the bombarded material they find stable configurations in normal lattice positions as well as in interstitial ones. In a second type of reaction these impurities trap other interstitials or vacancies or are trapped themselves by clusters or dislocations. At higher temperatures most implants in metals end up as constituents of solutions or alloys. If the percentage of the impurity is high and exceeds the solubility limit in the host, precipitations of the impurity or of certain phases take place: for example, in Al bombarded with Cu, $CuAl_2$ phases precipitate[45], while rare gases in metals form gas bubbles due to the interaction of vacancy clusters with the gas atoms[46]. All the effects mentioned in this paragraph are of immediate importance for materials subjected to neutron or heavy-ion irradiation, as for example nuclear reactor or fusion reactor constituents.

3 The Experimental Implantation Technique

This chapter deals with methods and equipment used for the production and acceleration of ions, and for the interaction of the ions with the targets. A great number of review papers and conference proceedings on ion sources and low energy accelerators provide detailed information for everyone interested in the subject[47–49, 50].

The following pages give a short survey without going into details.

G. K. Wolf

3.1 Ion Accelerators

Every type of machine for the acceleration of heavy ions consists of a source for the production of ions, installations for a one-stage or multistage acceleration and for shaping and cleaning the beam, and a target chamber containing the target and the equipment for the beam diagnostics.

The most important apparatus for ion implantation is the low-energy accelerator producing ions with energies up to a few hundred keV. The acceleration is performed in one stage, the ion source being on high potential and the target section on ground potential. Sometimes the target is on negative potential, thereby providing a post-acceleration. In this case the total ion energy is the sum of the preacceleration and the postacceleration.

One type of accelerator widely in use is the so-called Scandinavian type. Fig. 15 shows schematically the functions and the potential distribution along this machine: The ions from the source are extracted through a round aperture by means of an extracting electrode on a lower potential than the ion source. The focusing is done with the extraction electrode and another electrostatic or quadrupole lense. The centre part of the apparatus is a magnet for cleaning the ion beam by isotope separation. There are different types of magnets on the market ranging from 30° up to 180°. Depending on the quality of the magnet, one may achieve a separation in a way that the contamination from a neighboring mass is less than 1 %. After separation the beam enters the target chamber with the diagnostic elements and the target holder (see below).

Another type of low-energy accelerator is the Harwell machine developed by H. Freeman. The main difference between it and the Scandinavian machine lies in the extraction geometry. The ion source has an exit slit instead of an exit hole. This allows much higher ion currents. Unfortunately, focusing with lenses is difficult in this case because of space charge limitations. Therefore, the shaping of the beam of the Harwell separator is done with the help of the fringing field of the magnet and ajustable pole faces at the entrance and the exit.

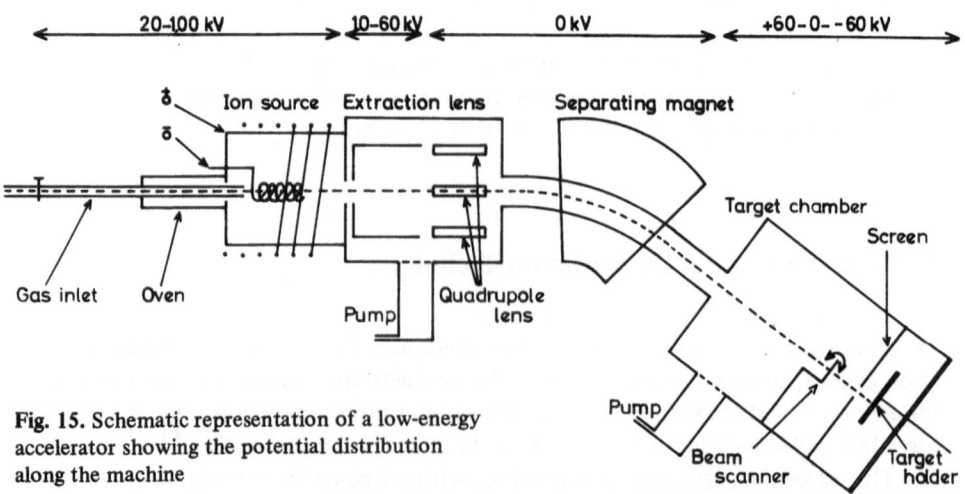

Fig. 15. Schematic representation of a low-energy accelerator showing the potential distribution along the machine

If energies above $\approx$ 1 MeV are desired, one has to use other types of accelerators, for example the ones in use for research in the field of atomic physics and nuclear physics. The van de Graff and the linear accelerators with multistage acceleration are especially popular. For a description of their function the reader should consult the literature on applied physics and particle accelerators.

To summarize, one may use the Scandinavian type accelerator for the production of ions of all elements with energies up to a few hundred keV and intensities of $\lesssim$ 10 μA (6 x 10^{13} ions/sec) or the Harwell type for intensities of $\lesssim$ 100 μA (6 x 10^{14} ions/sec). There are also a number of other constructions which differ to some extent from the types mentioned. Multistage accelerators, for instance, deliver ions with energies up to 10 MeV/A (MeV/nucleon) with intensities $\lesssim 10^{13}$ ions/sec.

3.2 Ion Sources

Until now we have not mentioned in detail the function of the ion source of an accelerator. There are numerous types of ion sources but only a small number is listed in this chapter. Most sources have in common that the element in question enters the source in gaseous form and is extracted from a hole or slit in the source by means of a potential difference. The "normal" electron impact sources known from mass spectrometers are inconvenient for ion accelerators because their ionization efficiency is too low (typically 10^{-2} %). The most important sources in use are:

3.2.1 The Radio-frequency Source

In this source the gas inside a tube is ionized by radio-frequency ($\approx$ 100 Watts, 10–100 MHz). A magnetic field concentrates the discharge on the extraction region. The extraction takes place due to a potential difference between the discharge and the beam extractor. The source works very conveniently for gases. For some applications it has the drawback that the energy spread of the ions produced is larger compared with other sources.

3.2.2 The Duoplasmatron

The ionization is achieved by a discharge between a cathode and an anode in which an exit hole is drilled. An intermediate electrode confines the discharge to a small region. The plasma extends through the anode aperture and the ions are extracted from the plasma surface by means of an extraction electrode.

3.2.3 The Surface-ionization Source

The operation principle of this source is very simple. When a beam of vaporized metal hits a hot surface such as Ta, W, Ir, the metal atoms are ionized due to an interaction with the surface. This may take place in a tube consisting of the ionizing metal.

An acceleration electrode extracts the ions from the end of the tube. The ionization efficiency η is given by the Saha-Langmuir formula[51]:

$$\frac{1}{\eta} = 1 + \omega \cdot e^{[e(I-\phi)/kT]}$$

ω = statistical weight ratio of atoms/ions adsorbed at the surface
I = ionization potential
ϕ = work function of the ionizer

Typical values for a substrate temperature of 2000°K are:

$\eta \approx 100\,\%$ for Cs, Rb, K
$\eta \approx\;\; 50\,\%$ for Na, Ba
$\eta \approx\;\;\; 1 - 10\,\%$ for Sr, In, Al, Ga

Thus surface ionization is a superior method for a restricted number of elements.

3.2.4 The Scandinavian (Nielsen) Source

The source sonsists of an anode cylinder enclosing the discharge region and a filament for the production of electrons which acts as the cathode. A magnetic field forces the electrons to move radially in order to lengthen their path. The ions are extracted from the plasma surface which extends through an aperture in the discharge chamber.

A number of sources are modifications or further developments of the Scandinavian-type source:

The hollow cathode source in which the plasma is contained inside the cathode (advantage: very high temperature).
The source developed by Kirchner[53] being much smaller than the Nielsen source and containing an intermediate grid, a modification which causes very low contamination from neighbouring masses.
The Harwell source having a rectangular extraction geometry and an axial arrangement of the filament near the outlet slit.
All these sources allow the production of ions of nearly all elements with efficiencies between ≈ 1 and $50\,\%$.

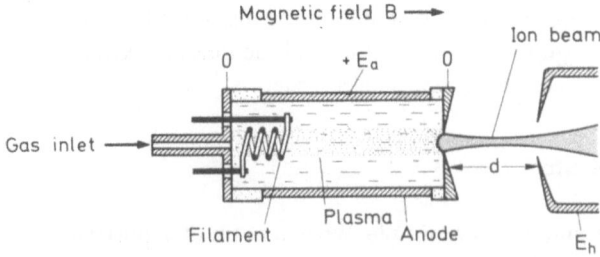

Fig. 16. Schematic view of a Nielsen source[52]

3.2.5 Methods for the Evaporation of Solid Materials

Most ion sources are fed with gaseous material. Thus, the initially solid material has to be vaporized before entering the ionization region. In most cases this may easily be done by using an oven directly coupled to the gas inlet of the source. Difficulties arise from elements with very high melting points such as, for example, platinum metals, or with sources having a low working temperature, where a condensation of the material inside the source may take place. There are different techniques to overcome these problems.

The temperature in a Scandinavian source is, for example, around 1000°C. If this is not enough for certain elements, one may use a hollow cathode source which operates at 1500 − 2000°C or even a special high-temperature source using electron bombardment for the vaporization of the material. In that way one may achieve temperatures as high as 3000° C[54].

Another possibility is the sputter ion source. Here inside the source gaseous ions are accelerated to the reflector electrode containing an insert made of the element to be vaporized. Due to the sputtering effect, the material evaporates and is converted into ions. This technique is universal and applicable for all elements. Fig. 17 shows a sketch of a sputter source[55].

If one does not have the possibility of using different sources for different elements, one may feed the source with compounds more volatile than the elements. Chlorides, fluorides, and oxides, in particular, are useful feed materials. Unfortunately, some compounds, which are otherwise suitable, tend to undergo thermal decomposition in the source and therefore have to be eliminated. Others, such as some chlorides, are extremely sensitive to water and hence converted under normal experimental conditions into their non-volatile oxides. To overcome this difficulty, the CCl_4 method was developed. With it the desired element is stored as an oxide in a rather hot zone near the inlet of the ion source. By passing CCl_4 over the oxide an "in situ" chlorination takes place and the resulting chloride is immediately swept in-

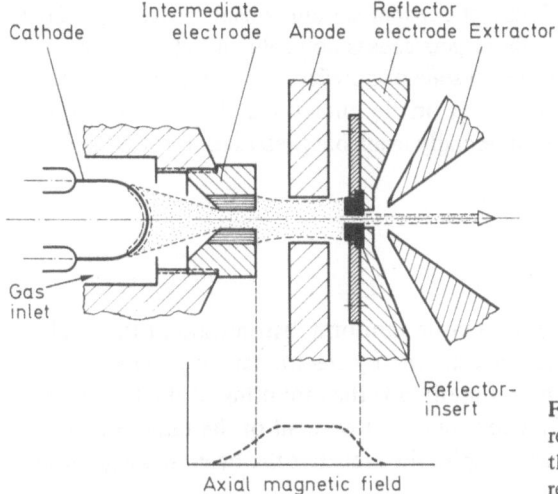

Fig. 17. Duopigatron ion source with reflector electrode insert for sputtering the desired element into the ionization region

to the discharge chamber[56]. If one uses compounds as source material, the formation of a considerable fraction of molecular ions has to be taken into account, thus complicating the mass spectrum and the formation of clean beams of distinct ions.

A compilation of evaporation methods for the different elements has been published by Freeman and Sidenius[57, 58].

3.3 Target Techniques

Techniques for handling the targets inside the accelerator and for observing the beam and the dose received by the target are mentioned under this heading. A comprehensive description is available in[47]. The problems connected with the interaction of ion beams with gases will be treated in Sect. 4.4.2.

3.3.1 The Mounting of Solid Targets

There exist numerous constructions, but two types are mainly in use: In one case 10 – 20 targets are mounted on a metal disc which may be rotated in a such way that every target is hit by the beam continuously or batchwise. Thus one gets either a greater number of targets with the same irradiation energy and dose or each target with different dose and energy values. The disc may be connected via a copper rod, fed through the walls of the target chamber to a container with liquid nitrogen for cooling the targets or a heater for high temperature irradiations. In the second case the targets are mounted on a roundabout which allows horizontal movement and the same irradiation cycles as the disc. Fig. 18 shows a disc target holder.

For certain experiments an alignment of the target with respect to the beam axis is indispensable, in particular for sputtering as a function of angle measurements or channeling investigations with single crystals. For that purpose several goniometers have been designed and some are on the market which one may operate inside the vacuum chamber. For further details the reader should consult books on chaneling[59, 47]. While the mounting of metal targets causes no problems, chemical compounds require some treatment. Either crystals or powders pressed into pellets have to be used. Sometimes these targets are very brittle, then it may be favorable to press the powder into metal crucibles. Commercially available presses used for IR spectroscopy are convenient tools.

3.3.2 Uniform Beams and Profiles

For a number of applications one needs a rather uniform distribution of the implanted ions over the target area. This may be achieved by electrostatically sweeping the ion beam over the target. An even better method is the mounting of the target on movable frames or rotating discs. The continuous movement of the target in two directions leads to a uniform coverage despite the fact that the beam has a gaussian distribution in the X and Y direction.

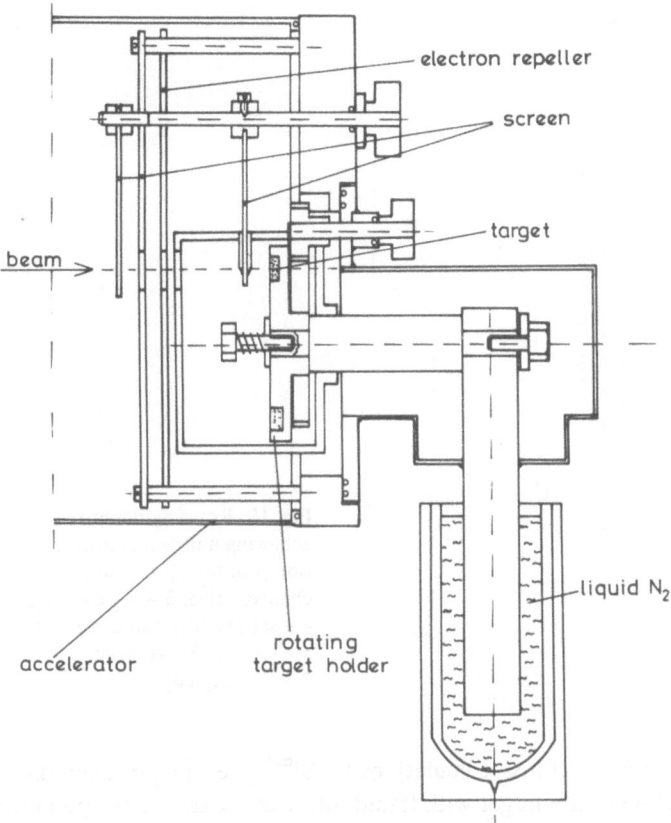

Fig. 18. Target holder for operation at liquid N_2 temperatures. The targets are mounted on a rotating disc.

Another problem is the shaping of the beam profile below the surface of the target. We have already mentioned the methods of sputtering and of implanting a sample with different energies. Besides these methods, one has the possibility of bombarding rotating target rods. In this way the beam hits the target under continuously changing angles and the depth distribution is relatively uniform from the surface to the mean range. Fig. 19 shows a rotating target[60].

To achieve a very low penetration depth, for example for the production of thin films or surface layers, the application of a retarding potential is helpful. A voltage of + 39 kV at the target and an acceleration voltage of + 40 kV results in a beam energy at the target surface of 1 keV and correspondingly a very small range.

3.3.3 Target Temperature

The ion beams generally in use may cause a considerable heating of the target. If one wants to bombard under controlled temperature conditions, one has to have a feeling for the extent of this heating. A beam with an energy of 100 keV and an intensity of 1 $\mu A/cm^2$ (6×10^{12} ions/cm^2/sec) corresponds to 0.1 watt of power.

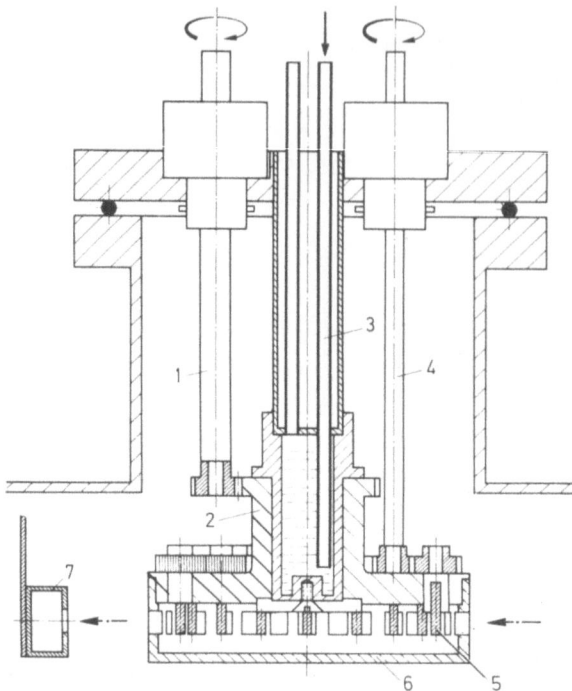

Fig. 19. Rotating target for achieving uniform concentration profiles. 1, 2 = sample changer drive, 3 = N_2 cooling, 4 = sample rotation drive, 5 = sample, 6 = Al cage, 7 = Faraday cup

Assuming only a radiative heat loss, calculations for $Si^{61)}$ give a target temperature of $50 - 100°C$. Cooling of the target with liquid nitrogen lowers the temperature by $\approx 50°C$. In practice, these are upper values because of the contribution of heat conduction. This effect is naturally more important for metals than for insulators. In general, if one uses high beam current densities, one has to take care of effective cooling, good heat contact or dissipation of the beam energy by means of defocusing or target movement.

3.3.4 Beam Diagnosis

During an implantation experiment one needs information on the beam profile, the focusing and the beam current or the dose received by the target. The beam profile and shape may be made visible with a screen. The most simple construction is a metal plate covered with KBr. A more elaborate system is a profile grid with an oscilloscope display. On the oscilloscope one sees the intensity distribution of the beam in the X and Y direction. The most convenient apparatus for low energy accelerators is a beam scanner. A needle moves perpendicularly to the beam direction and the current is displayed on an oscilloscope as a function of the position of the needle. The display gives information on the beam shape and on the neighbouring masses. Thus it is possible to control the focus as well as the quality of the mass separation. By using two needles moving in the X and Y direction one gets a complete picture of the beam.

Fig. 20 shows a display of a separated tin beam.

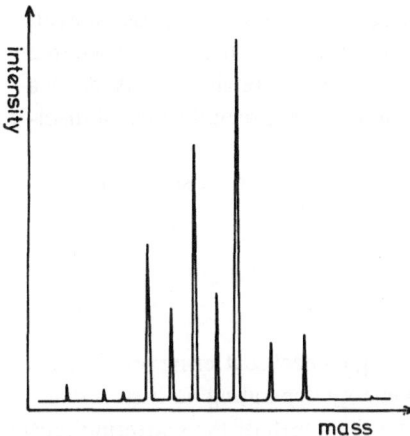

Fig. 20. Spectrum of tin isotopes as displayed from the target section of a low energy ion accelerator (after[62])

The measurement of the beam current hitting the target is in principle not difficult. If the target is a metal, one measures the current directly on the target. Integrated it gives the total dose[63]. Since a target hit by a beam emits secondary electrons, one has to correct the measured values. Either the target has to be mounted directly inside a Faraday cup or the target current has to be compared with measurements of a separate Faraday cup which one has to move occasionally into the beam instead of the target. The same technique may be employed during the bombardment of insulators. Here a direct precise measurement is impossible because of the target being charged by the beam. Thick insulators are sometimes charged to such an extent that the incoming beam itself is defocused or deflected. In those cases one either has to cover the target with a conducting thin film or a grid or to bombard it with electrons from a hot filament in order to neutralize the target surface.

4 The Analysis of Irradiation Effects and Products

In analysing targets bombarded with ions one is as much interested in the defect structure of the target as in the physical and chemical state of the implanted ions and their spatial distribution. In this chapter a survey of the most important methods of analysis is given with special emphasis on the techniques useful for chemical investigations.

4.1 The Analysis of the Defect Structure of Ion-Bombarded Targets

4.1.1 Microscopy

Electron microscopy is one of the most powerful methods for studying the damage structure of irradiated materials. Its limitation lies in the observable size of the defects. While it is not possible to observe point defects, all types of extended defects with a diameter of $\gtrsim 20$ Å may be made visible. Therefore, this method is used mainly for the investigation of the size and density of clusters in metals as voids, dislocations and grain boundaries. Impurity clusters in the form of precipitations of insoluble elements or new phases may also be studied[64].

Finally, changes in the surface topography of targets caused by sputtering are easily observed.

4.1.2 Ion Backscattering

Fast ions interacting with a solid surface may undergo a backscattering process. They leave the target in a backward direction, with an energy depending on their initial energy, the mass of the scattering target atoms and the depth of the scattering centre below the target surface. Usually a beam of α-particles or protons of $1-2$ MeV energy is used, the target is mounted on a goniometer, and the scattered particles are analysed by means of a solid state detector.

As a result one gets the intensity of the backscattered particles as a function of their energy. By applying the formula for the energy loss of ions in matter one converts this into a "depth below the surface scale" for a homogeneous target. If the target contains more than one element, the situation is more complicated because of

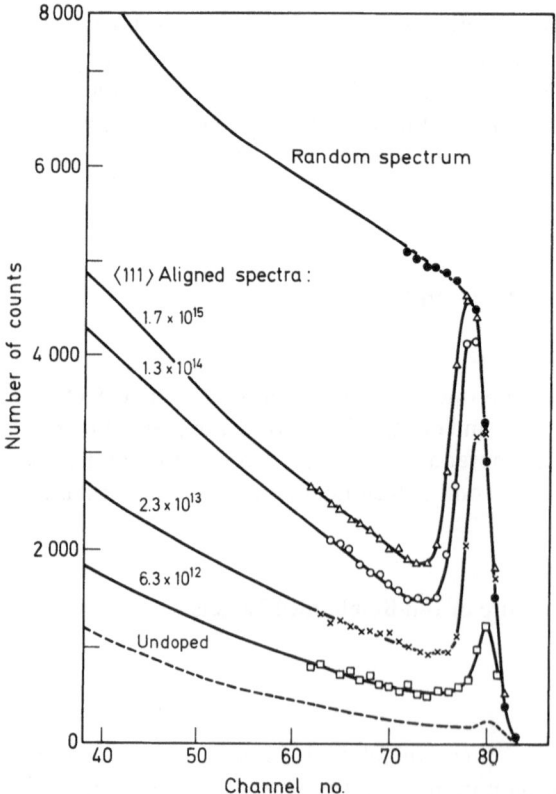

Fig. 21. Backscattering spectra from Ge crystals bombarded with In$^+$ ions. α-particle irradiation in aligned and random geometry[66]

the difference in energy loss caused by target atoms of different mass (Sects. 4.2.1 and 4.3.1).

With this method one is able to look to a depth of a few thousand Ångstroms into the material. Therefore, it is very well suited to yield information on the implantation process. The principle of studying damage is the following: A single crystal of the material in question is exposed to the α-particle beam in arbitrary orientation. Afterwards, the crystal is oriented to the beam axis in such a way that the α-particles are able to travel along channels in the crystal (channeling). In the first case, one gets a random backscattering spectrum, in the second one an aligned spectrum. The intensity of the backscattered α-particles as a function of the depth is much smaller in the latter case because the α-particles penetrate deeper into the crystal. Radiation damage now causes deviations from the ideal crystal structure, the number of available channels decreases, and due to interstitial atoms acting as additional scattering centres, the α-particle intensity becomes higher compared with the undamaged crystal. Fig. 21 illustrates this method. It shows the situation for undoped crystals in comparison with ones implanted with a high dose of ions. The peak contains the region damaged by the bombardment but the region below (low channel number) is not damaged and exhibits channeling. At a still higher dose the "damage peak" may reach the random spectrum, indicating that the damaged region is almost amorphous. The area under the damage peak is proportional to the number of displaced atoms. Further information on this method is given below and in numerous other articles[67 − 69].

4.1.3 ESR and Optical Spectroscopy

There are a number of spectroscopic methods for the investigation of defects in materials. In principle, one is able to apply all of them to the effects caused by ion bombardment. But since the penetration depth of the ions is only a small fraction of the total crystal, one has to take care that the effect of interest is intense enough in comparison with the background.

ESR or EPR belong to the techniques most frequently used. With them one is able to identify even rather low concentrations of defects. Typical ESR signals correspond to certain types of damage and one may follow the variation of the signal as a function of dose or annealing temperature. Many studies concerning the transition of the crystalline to the amorphous state have been performed with silicon in particular[70 − 72]. ESR is also a useful method for the identification of radicals in compounds. Problems arise mostly if one tries to attribute a certain signal to a well-defined type of defect.

Optical spectroscopy is another common method for the detection of defects especially in chemical compounds. The F, H, V centres, well known in ionic crystals, are easy to detect optically[73]. Because of the very low penetration depth of the bombarding ions the optical techniques have to be modified for implantation studies in the direction of reflection spectroscopy[74]. Finally, Raman scattering has proved to be of value for studying lattice disorder induced by ion bombardment in various materials[75 − 77].

4.1.4 X-ray Diffraction and Mössbauer Spectroscopy

Using X-ray diffraction general radiation damage is easily observable. In the special case of ion implantation the small penetration compared with the total thickness of the target causes problems. One either has to use the technique of thin layer X-ray diffraction where the information depth is $< 10^4$ Å or to restrict the investigation to ion-bombarded thin films.

The Mössbauer effect is also capable of yielding information about radiation damage. Here it is not possible to measure the damage directly but only the surrounding of an implanted Mössbauer isotope as $^{57}Co/^{57}Fe$, ^{83}Kr or $^{133}Xe/^{133}Cs$. Thus, the implanted ions act as probe for the formation and variation of the defect structure. The hyperfine interaction of the implanted atoms and the Mössbauer recoil less fraction in particular depend on the presence of defects. In [78] several suitable systems are listed. In Sect.4.2.4 a more detailed outline of Mössbauer measurements of bombarded targets is presented.

4.1.5 Measurements of Charge Carriers

The point defects created during ion irradiation cause an additional conductivity. The measurement of the resistivity is, therefore, a direct indication of the presence and the amount of point defects. This method has yielded important results on metals and semiconductors. The majority of "recovery" curves (annealing of point defects created at liquid He temperature) derives from resistivity measurements which are not too difficult to perform over a wide temperature range. Another possibility for following the recovery of defects created at low temperatures is the measurement of the release of the energy stored in defects using calorimetric methods.

Finally, the change in length and volume of a test specimen tells something about the vacancies or vacancy clusters present (swelling). One will find more information on this method in [79].

4.2 The Location and Chemical State of Implanted Atoms

In addition to the damage configuration of the bombarded target the location of the implanted atoms in the host and their chemical state is the most interesting question one would like to have answered. A number of methods, some of them already mentioned in the preceding section, yield information on this subject.

4.2.1 Ion Backscattering

This technique, which has already been described, may serve with some modifications for locating foreign atoms in a lattice:

We have seen interstitials causing an increase in the backscattering yield of ions when a single crystal is oriented in a such way that the particle direction corresponds

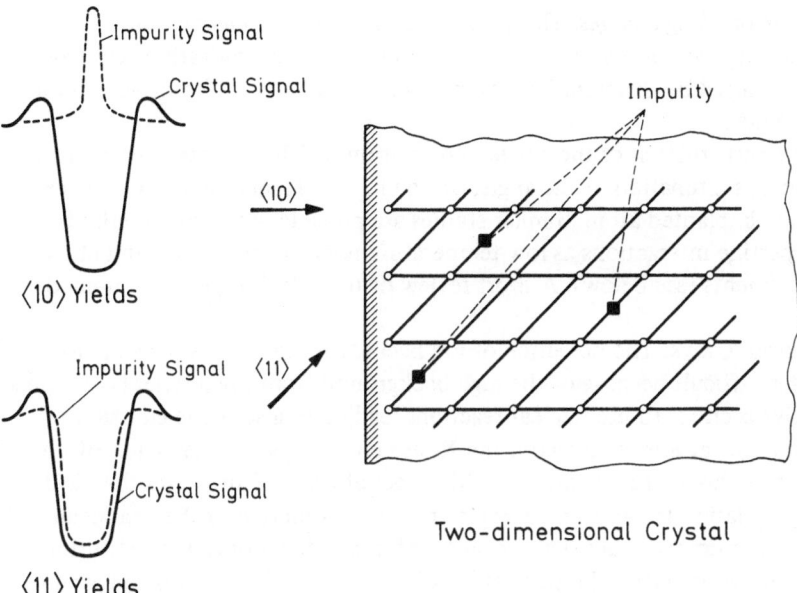

Fig. 22. A two-dimensional crystal containing impurity atoms (on the right) and the backscattering signal resulting from tilting the crystal around the ⟨10⟩ and ⟨11⟩ axis[80]

to a crystallographic channel. On the other hand, an implanted atom in a substitutional position does not lead to increased backscattering. But one has to be careful because no increase in backscattering yield does not necessarily mean a substitutional position but rather a site being shadowed by the normal row of atoms. An exact analysis of the foreign atom position requires experiments in different orientations. Fig. 22 shows the situation for a two dimensional model. Position ■ for example is not substitutional but shadowed in the ⟨11⟩ direction.

Experiments mostly involve tilting of the crystal around the crystallographic orientation in question and measuring the backscattering yield as a function of the angle. This directional effect for the implanted sample is compared with an unimplanted crystal.

Fig. 22, left side, shows as an example an angular scan of the ⟨10⟩ and ⟨11⟩ direction of our model.

The dip resulting from the implanted sample is shallower compared with that of the pure sample, indicating a small fraction of the impurity atoms not being shadowed. In the ⟨10⟩ direction we get even an increased yield in comparison with the random orientation, this being a sign for a position almost in the middle of the channel. Details of this technique and its application may be found in [81].

4.2.2 Perturbed Angular Correlation (PAC) and Ion-induced X-Rays

Alternative methods for lattice location studies are:

Perturbed angular correlation measurements: The principle relies on the fact that hyperfine interactions occur between nuclear moments and the electromagnetic field of the surrounding charges. The γ-quanta emitted in a nuclear cascade very often show an angular correlation. This correlation is affected by various electronic environments caused by different ligands in compounds or by different neighboring positions in solids.

From the perturbation of the angular correlation which one measures with two γ-ray detectors as a function of the angle, one deduces information on the lattice position of the implanted atom. Similar studies are possible with other methods based on hyperfine interactions as low-temperature nuclear orientation and Mössbauer measurements (see below). A short review of these techniques is contained in [82].

Ion-induced X-rays: The detection of ion backscattering signals from light implanted ions is difficult because of the high background from the heavier host atoms. The alternative is either to use nuclear reactions leading to a selective emission of particles from light atoms or to record the X-rays emitted as a consequence of the interaction of the bombarding particles with target atoms. In both cases the yield of secondary radiation for an aligned single crystal is a function of the channeling efficiency of the primary beam. That means that implanted atoms in substitutional or interstitial positions cause the same effects in the case of X-ray emission as in the one of heavy ion backscattering[83, 84].

4.2.3 Spectroscopic Methods

The classical methods for the analysis of the chemical state of an atom are the spectroscopic ones. Therefore, it is natural to consider them for studies of the chemical state of implanted ions as well. But in contrast to the investigation of radiation damage caused by the bombarding particles, one has the additional factor of intensity in the detection of the implanted ions themselves.

Therefore, UV spectroscopy and Raman and IR analysis are restricted to high dose bombardments where the concentration of the species formed is high enough to yield useful signals[85, 86a]. This is mostly the case for irradiations with $\geqslant 10^{16} - 10^{17}$ ions/cm^2.

ESR spectroscopy, on the other hand, is much more sensitive and may be also used for the analysis of low-dose irradiation products. It yields details about the defect structure around an impurity atom and the chemical nature of its neighbors. The only drawback of the method is the difficulty in interpreting the ESR spectra unambiguously[86].

4.2.4 Mössbauer Spectroscopy

The technique of Mössbauer spectroscopy is especially suited to chemical studies of implanted ions. The great amount of reference material allows in most cases rather detailed statements on the chemical bonding of an element implanted in compounds. The implantation dose necessary for Mössbauer spectra is not too high. That means that the lattice order is normally preserved.

On the other hand, Mössbauer spectroscopy is not a universal method, but restricted to the elements for which suitable Mössbauer nuclides exist. In the conventional type of experimental arrangement a standard radiation source, $^{57}Co/^{57}Fe$ for example, is used, and the solid of interest serves as an absorber. The energy difference between emitted and absorbed radiation is measured using the Doppler shift method. From the resulting spectra one extracts knowledge about the electron density around the Mössbauer nucleus, and the symmetry and the electron distribution of the ligands[87, 88]. This type of experiment is not appropriate for the investigation of ion-bombarded samples because it requires milligram quantities of the compound under study.

Another possibility is emission spectroscopy. Here the sample under study serves as a source and is measured against a standard absorber. Applying this method to ion implantation naturally requires the bombardment of the sample with radioactive ions. From such radioactive implanted hosts as Mössbauer sources one may learn something about the defect structure as well as about the lattice location and the chemical state of the implanted element[78]. In the first case, already mentioned in the preceding section, the implanted ion serves only as a probe for monitoring the surrounding defect; in the second case the question of its substitutional or interstitial position is deduced from the symmetry of the immediate surroundings. Finally, the Mössbauer spectrum tells us the oxidation state and the bonding conditions.

Very interesting results may be obtained with a Mössbauer spectrometer run directly on-line with an ion implanter[89, 90].

This arrangement allows bombardments at liquid He temperatures with subsequent measurements without change of temperature or time-consuming target transport. This is very important because very often the majority of the annealing steps take place below room temperature. Fig. 23 shows a set-up[89] consisting of the target chamber of the accelerator, the liquid He cryostat with the Mössbauer source being simultaneously the implanted target, the Mössbauer drive and the reference absorber. The lowest temperature achieved during implantation is $15°$ K.

Emission spectroscopy has some serious drawbacks:

The implantation of millicurie quantities of radioactive elements is only possible in "hot laboratories". If the radionuclides are long-lived, such as ^{57}Co with a half-life of 270 d, one contaminates the accelerator for years.

The Mössbauer emission spectrum reproduces two effects: the situation of the implanted atom in the host and the so-called after-effect. The latter expression means that, for example, in the case of ^{57}Co having been implanted, we first get the radioactive decay of $^{57}Co \overset{\epsilon}{\to} {}^{57}Fe$ and then the Mössbauer transition to the ^{57}Fe ground state. The electron capture or β-decay itself very often causes a perturbation of the environment of the isotope in question[91]. Thus some knowledge about the pure after-effect is needed for the interpretation of the spectra of implanted samples.

Recently a new technique has come into use which avoids radioactivity as well as after-effects: conversion electron spectroscopy[92, 93]. This method uses the fact that in most Mössbauer transitions not only γ-rays but also conversion electrons are emitted. In the case of ^{57}Co the electron conversion is the major decay mode. Thus, instead of measuring the γ-rays, absorbed or "reflected" from an absorber, one measures the emission of electrons from the absorber as a function of the velocity

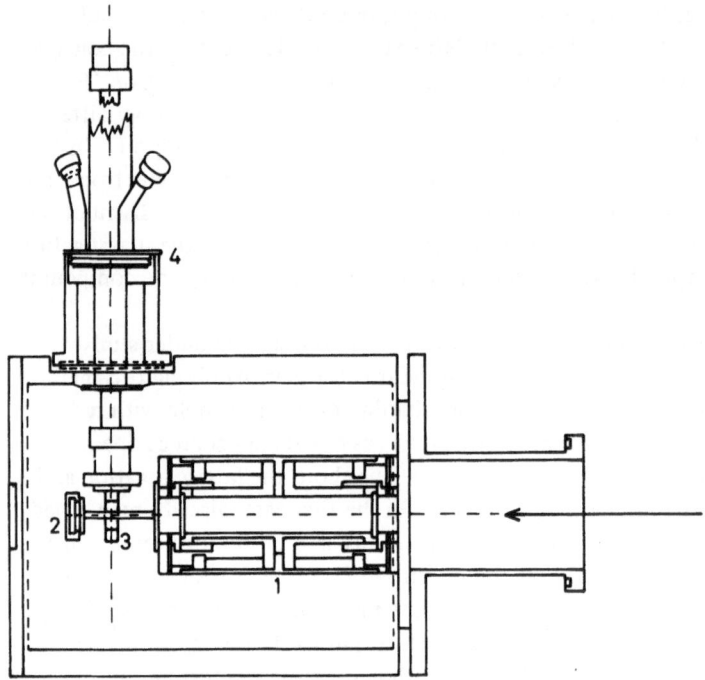

Fig. 23. A Mössbauer spectrometer coupled to an ion implanter. 1 = transducer, 2 = absorber 3 = source, 4 = He flow cryostat. The beam comes from the right, the detector is situated at the left

of a standard source (Doppler effect). Since the depth from which the electrons are emitted without self-absorption lies exactly in the range of the penetration depth of the bombarding ions, the method is sensitive to the state of the implanted region without any interference from the unimplanted layers further below the surface. The possibility of measuring conversion electrons from the K and the L shell independently allows getting information about different layers below the surface because of the difference in energy and, correspondingly, in range between K and L electrons.

4.2.5 Chemical Analysis

In addition to the physical methods mentioned above there exist chemical analysis techniques well known from the field of hot-atom chemistry[94 − 96]. The application of these techniques to ion-bombarded samples requires the implantation of radioactive ions. The principle of the method is the following:

The irradiated samples are usually dissolved in a suitable organic or inorganic solvent. Afterwards selective precipitation, solvent extraction, ion exchange or chromatographic procedures take place. By counting the radioactivity one is able to localize the fractions containing the implanted atoms and their reactions products. Very often macroscopic amounts of the reaction products expected have to be introduced into the solvent beforehand to permit an exact separation. A typical example is the famous experiment of Szilard and Chalmers[97] who found that energetic radioactive recoiling iodine atoms produced and slowed down in ethyl iodide are partly extractable with organic solvents and partly with aqueous solutions. From these findings one may draw the conclusion that the recoiling iodine is partly bound to organic residues (mainly C_2H_5) and partly present as "inorganic" I^-. A comparable procedure is the evaporation of the bombarded sample or the resulting products. Radioactivity in the volatile fraction or in the residue tells something about the volatility and vapor pressure of a chemical compound formed from the implanted atom and the target atoms or molecules[98, 99].

The advantage of this type of analysis is the great number of separation and identification methods available by using numerous procedures from analytical chemistry. Unfortunately, it is a destructive method and the products alter because of the high probability of reactions taking place with the solvent during dissolution. Thus the results one gets are only indirect ones and allow conclusions about the conditions in the solid only to a certain limit.

4.3 The Analysis of Surfaces and Profiles

Knowledge of the composition of the immediate surface of ion-bombarded samples is of interest for various reasons:
— the sputtering effect might change the composition of the surface,
— the equilibrium between sputtering and implantation limits the maximum concentration of ions implantable in every system,
— radiation enhanced and thermally induced diffusion leads to an increase or decrease of the bombarding particles on the surface relative to the region below the surface,
— ion-bombarded surfaces are very reactive. Frequently one wants to know the oxygen, nitrogen and carbon content of the surface,
— finally, every surface analysis technique allows, together with the removal of thin layers, the determination of the depth distribution of the implanted impurities.

In this section the most important surface analysis methods which are effective to a depth of $\lesssim 50$ Å are mentioned. They are based either on the interaction of ions with surface layers, the ejection and analysis of electrons from the surface, or the investigation of secondary ions.

4.3.1 The Interaction of Ions with Surface Layers

The basic technique has already been mentioned in preceding sections. The a- or proton-backscattering method has to be modified so that the recording of light surface impurities such as oxygen or nitrogen on a heavy target is allowed. Normally the light element peaks in the backscattering spectrum are hidden by the heavy element peaks. The situation is much better if one uses thin films of a substrate on a light element backing. Under these conditions N, C and O are also detectable. Another method for improving the resolution for the surface region is the glancing-incidence method. Here the incoming beam hits the target at an angle of $\geq 170°$ and penetrates only into the surface layers of the sample. This method is certainly more sensitive to the surface composition than normal backscattering measurements. Finally, channeling of the particles along an aligned single crystal depresses the yield from the bulk atoms and enhances the yield from the disordered surface atoms[100, 101, 83].

Instead of particles backscattering from the target one may also use particles ejected during nuclear reactions for the surface analysis. The advantage is that the nuclear reaction is specific for certain light elements:

For example: ^{12}C (d, p) for Carbon
^{16}O (d, p) for Oxygen
^{18}O (p, α)

The deuteron or proton beam induces the nuclear reaction and the energy of the outcoming protons or α-particles depends on the depth and concentration of the impurity. These particles may be distinguished from elastically-scattered primary particles due to their energy[83].

The analysis of ion-induced X-rays is based on the same principle. The incoming particles generate characteristic X-rays and their analysis allows the evaluation of concentrations of certain elements. The sampling depth is > 100 Å. Therefore, it is not really a method for shallow surface layers[102, 103]. On the other hand, since the cross section for the generation of soft X-rays is rather high, this can be used for analysing layers nearer to the surface[104].

4.3.2 The Ejection and Analysis of Electrons from the Surface

During bombardment of solids with electrons or X-rays secondary electrons from the target atoms are excited and partly ejected. The ejection of K and L electrons is followed by a reordering process which leads to the emission of X-rays or Auger electrons from outer shells.

The analysis of the electrons is done according to energy and intensity. Small shifts in the energy give information about the chemical bonds, but the interpretation of the results is difficult. The main use is therefore the investigation of the composition of the surface of samples, this being possible by means of comparison with standards of known composition. The method of analysing the K or L electrons is known as ESCA (Electron Spectroscopy for Chemical Analysis)[105]. The sampling depth is $30 - 100$ Å depending on the element. More specific for the immediate surface is the method of Auger spectroscopy, analysing the Auger electrons. Due to their lower energy the sampling depth is only $5 - 20$ Å[83, 106, 107].

4.3.3 The Ejection and Analysis of Ions from the Surface

Due to the sputtering process an incoming ion beam ejects secondary atoms and ions from the surface. The secondary ions may be analysed by means of a mass spectrometer (SIMS).

The mass spectrometer gives a picture of the surface composition. It is sensitive for hydrogen as well as for uranium. There are two modes of operation: in the first case one works with a primary beam of very low intensity in order to avoid changing the surface composition extensively. This then gives spectra of the first few layers of the sample. In the second case, one erodes the sample to a greater extent in order to get spectra of a layer of 100 – 1000 Å below the surface.

Quantitative analysis is not easily achieved for various reasons:
- sputtering of surface contaminants (hydrocarbons from the pumps),
- the relative percentage of ions compared to atoms is quite low and varies from element to element,
- during the sputtering process the composition of the surface of the sample changes very often due to preferential sputtering of certain elements.

Some of these difficulties may be overcome by employing comparative well-pre-characterized standards. Nowadays SIMS has developed to an important technique in surface analysis[108 – 110].

4.3.4 The Measurement of Concentration Profiles

Very often the knowledge of the composition of the surface or of the bulk of a specimen is not enough, because one is interested in the concentration of a certain element, the implanted atom for example, as a function of the depth below the surface.

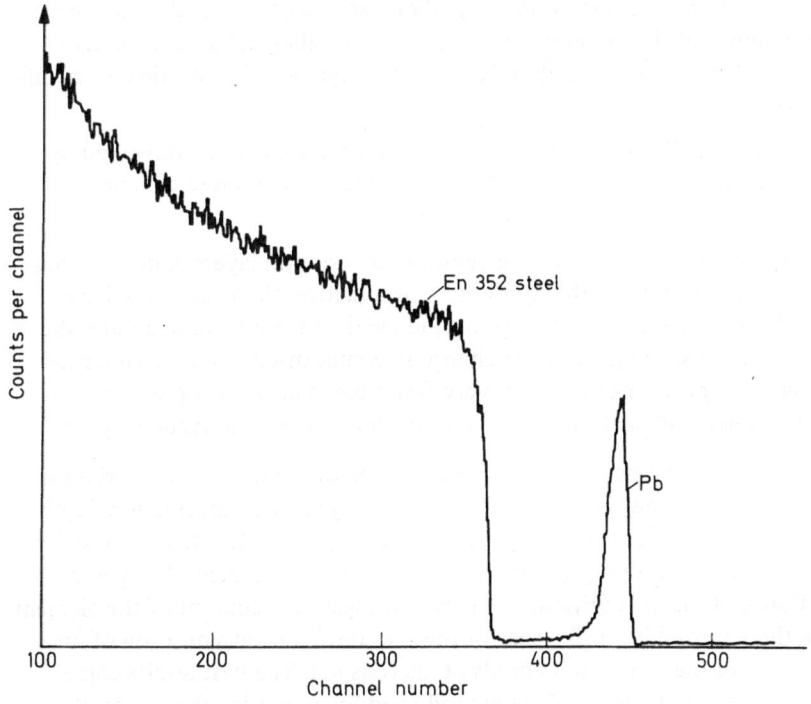

Fig. 24. α-backscattering spectrum from steel implanted with Pb^+ [112]

There exist two types of techniques, a non-destructive and a destructive one:

a. The Non-destructive Method[111]. The backscattering technique is the most important method for non-destructive measurements. The energy of the backscattered particle tells us something about the mass of the scattering centre (impurity) as well as the depth. With help of the stopping power formulas the backscattering spectrum may be transformed into a relationship between concentration and depth for a distinct element. In that way it is no problem to evaluate the range profile of implanted ions having a higher mass than the target. Fig. 24 shows as an example a spectrum of lead in stainless steel.

The lead peak on the high-energy side of the spectrum reflects the lead profile in the steel sample. For light elements range profiles are only measurable in single crystals with the help of the channeling effect or in polycrystalline material using specific nuclear resonance reactions (see preceding section). Another possibility for single crystals offers the measurement of the yield of ion-induced X-rays excited by channeled energetic ions[84].

b. The Destructive Method[108, 111]. The majority of methods for achieving range profiles are based on a combination of sectioning of the sample and analysis of the change in the surface composition. The removal of thin layers of only a few Å thickness is not easy.

The techniques mostly used are:

Chemical or electrolytic dissolution. The disadvantage of this method is the non-uniform removal of the target layers. Different crystallographic faces, local structures and small regions with different electrolytical behavior dissolve at different rates.

Anodic oxidation. The specimen is anodized to a certain extent controlled by current potential conditions. Then the anodized layer is removed chemically or mechanically.

Sputtering. Theoretically the sputter removal of very thin layers under controlled conditions is possible with any element. In practice, there arise problems because of the dependency of the sputtering yield on the crystal structure, the orientation of the specimen and the change in composition due to preferential sputtering. The sputtering method is very favorable in cases of subsequent removal of material and analysis of the composition in the same vacuum system.

For the analysis of the new surface after every removal one may use all the surface techniques already mentioned in Sect. 4.3.1 as long as their information depth does not exceed the thickness of the layer removed: Auger and ESCA-spectroscopy, secondary-ion mass spectrometry (SIMS), backscattering, ion-induced X-ray and nuclear reaction analysis. In addition, one may investigate the content of the element of interest in the removed layer. Because of the low absolute concentration of implanted ions most of the standard methods of analysis fail. The best results come from implantations of radioactive elements followed by measuring the radioactivity of the dissolved removed layer.

4.4 Special Methods

4.4.1 The Microprobe

It is often necessary to know not only the composition of a sample but the variation in composition along the x and y direction as well. For this purpose one needs a well-focused analysing beam which is scanned across the surface of the specimen. This so-called microprobe utilizes either an electron or an ion beam to create X-rays or secondary ions measured in the way already mentioned in Sect. 4.3.3. The lateral resolution of the system is typically in the order of 1 μm, the sampling depth from 10 Å to some μm[83].

4.4.2 The Detection of Gaseous Implantation Products

During ion bombardment gaseous reaction products may be formed on or below the surface of a sample. The reason may be a catalytic reaction involving impurities absorbed at the surface or the interaction of implanted atoms with each other or with target constituents.

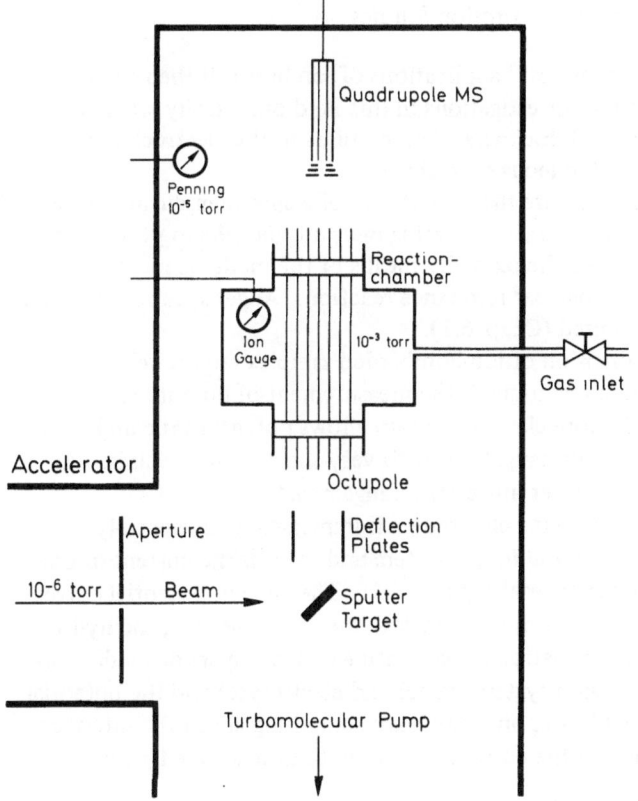

Fig. 25. Apparatus for the study of products released from a sputter target

For example:
$$x \cdot C^+ + y \cdot H^+ \to C_x H_y ;$$
$$x \cdot H^+ + y \cdot FeC \ (Target) \to C_y H_x \ in \ FeC$$

By sputtering or heating of the sample these reaction products may evaporate into the gas phase. During this process decomposition induced by thermal reactions or the sputtering beam may take place, but this is by no means always the case. Even rather complicated molecules may leave the target without decomposition. A suitable method for recording the atoms or molecules released is mass spectroscopy. Mass spectroscopy is extremely sensitive to the release of ions, less sensitive to neutrals because of the ionization efficiency (typically 0.01%) of the ion source of the spectrometer.

Such studies are possible with a SIMS spectrometer or a similar arrangement. Fig. 25 shows the set-up used in our laboratories[60, 113]. It consists of an ion accelerator and a target subjected to ion bombardment in order to implant ions or sputter the surface. The gaseous products are removed from the surface by sputtering or heating and the resulting ions are guided by an octupole to a quadrupole mass spectrometer prior to analysis. Besides this the interaction of atoms with a gas in the reaction chamber may be studied. The arrangement allows the identification of gaseous reaction products, the recording of release curves (gas release against temperature), and the measurement of depth profiles for gaseous atoms or molecules inside the target.

4.4.3 Methods and Equipment for Corrosion Studies

One of the most important "chemical" applications of ion beams is their use in corrosion science (see below). In investigations in this field one usually wants to know either the composition and thickness of oxide films or the electrochemical behavior of corroding samples (aqueous corrosion).

In the first case one may measure the weight gain of a specimen during oxidation. Unfortunately, this is a very time-consuming method (months to years). Another possibility is the analysis of the oxide by means of the methods mentioned in Sect. 4.2[114]. α-backscattering, nuclear resonance reactions, Auger spectroscopy and ESCA in particular have been used (Chap. 6.1).

The study of aqueous corrosion of ion-bombarded samples requires electrochemical equipment. The basic technique is the measurement of potentiostatic current-potential relations. A more elaborate set-up allows potentiostatic and galvanostatic measurements and also investigations with very short current pulses. Fig. 26 shows a schematic lay-out of a rather universal arrangement[115].

The sample studied is the working electrode AE immersed in an electrolyte. Either one applies a defined potential to the system and records the current or one defines a current and measures the resulting potential. The current-potential curves give information about the mechanism of corrosion, the corrosion rate, the hydrogen-evolution process and the formation of passivating surface layers depending on the pH and nature of the corrosion system (metal and electrolyte) and the potential or current range applied. In addition, one may learn something about the interface solid/liquid from the variation of the current density with time at fixed potentials[116 – 118].

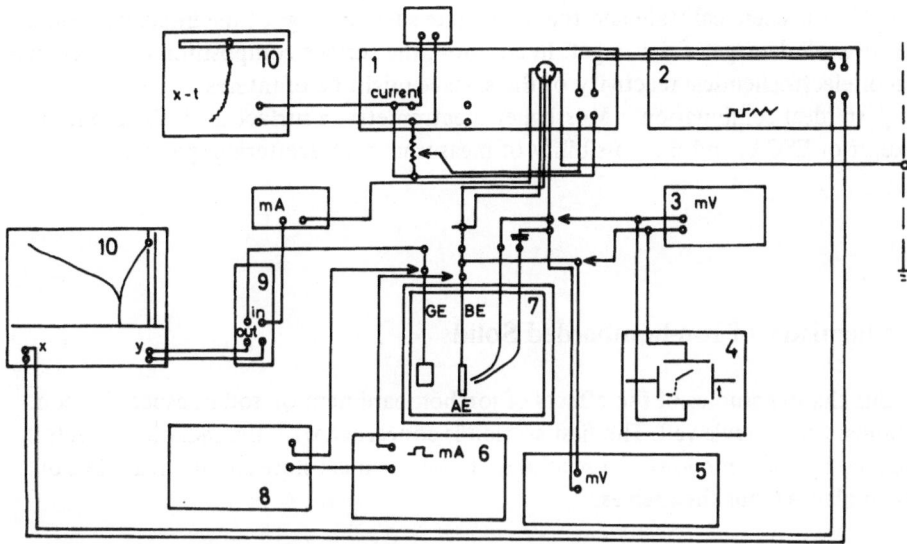

Fig. 26. Scheme of a set-up for measuring current-potential relationships.
1 = potentiostat, 2 = plotter regulation, 3 = digital voltmeter, 4 = oscilloscope, 5 = voltage regulation, 6 = galvanostat, 7 = electrolytic cell containing the working electrode AE, the reference electrode BE and the counter electrode GE, 8 = integrator, 9 = log. converter, 10 = plotter

4.5 Summary of the Procedures Most Important for Chemical Studies

The following table contains the methods which are particularly useful for the chemist studying the consequences of ion bombardment. In most cases one is inter-

Table 3. Procedures for the investigation of the chemical consequences of ion bombardment

	Problem	Analytical procedure
Surface	Composition	Auger-, ESCA-spectroscopy, Ion-beam analysis
	Chemical reactivity	Standard chemical surface techniques
	Electrochemical reactivity	Current-potential relationship
Bulk	Radiation damage	Ion-beam analysis, Optical spectroscopy, X-rays, Electron-spin resonance
	Lattice location. State of immediate surroundings of implant	Mössbauer spectroscopy, Electron-spin resonance, Ion-beam analysis
	Chemical state of implant	Mössbauer spectroscopy, Dissolution and analytical treatment, Electron-spin resonance

ested in the chemical state and the immediate surroundings of the implanted ion or the over all damage of the target. In addition, the surface composition or the chemical or electrochemical reactivity of the surface might be of interest.

An ideal combination is Mössbauer spectrometry, a surface analysis technique (Auger or ESCA) and the possibility of measuring back-scattering spectra.

5 Chemistry of Ion-bombarded Solids

In this chapter studies of the effects of ion bombardment on solids, especially compounds, will be reviewed. The first section is concerned with the radiation effects induced by energetic ions, the second with the chemical state and the reactions of the implanted ions themselves.

5.1 Radiation Chemistry of Solids Irradiated with Heavy Ions

While a lot of work has been done with heavy-ion induced radiation damage in metals and semiconductors, very little has been done on compounds, especially on covalent ones. In the following section we shall try to cover the most important work on the radiation chemistry of compounds. In the section on metals only a few selected examples will be mentioned. In the last part of this chapter we shall deal with changes in the composition of surfaces due to sputtering of compounds.

5.1.1 Radiation Damage in Compounds

Contrary to metals where radiation damage is only due to the displacement of atoms, in compounds radiation effects involving the electrons of the target also have to be taken into account. That means that the electronic stopping power plays an important role. In ionic crystals electronic processes lead to the formation of "centres" as vacancies containing electrons or oxidized and reduced species in lattice or interstitial positions. In covalent compounds the rupture of bonds is the dominating process. Since energy transfer is a significant parameter in these cases the effective interaction radius of an implanted ion might be larger than the region of the displacement cascade.

The number of displacements (N) from the nuclear interaction may be estimated using the formula

$$N = \frac{E_{pr}}{2\,Ed} \quad (E_0 \gg Ed) \qquad \begin{aligned} E_{pr} &= \text{energy of the primary particle} \\ Ed &= \text{minimum displacement energy} \end{aligned}$$

For Al_2O_3 (irradiated with 100 keV Pb^+), for example, the estimation yields a dose of 7×10^{13} ions/cm^2 for displacing every atom in the lattice once. The result should be an amorphization of the bombarded region. The comparison with experimental data shows that $> 10^{15}$ ions/cm^2 are needed to obtain the transition from the crystalline to the amorphous state. The main reason for this discrepancy is the spontaneous recombinations of defects during the irradiation. 100 % nuclear stopping of the Pb$^+$ ions was assumed for the above estimate. A different approach would be the assumption of 100 % electronic stopping. Since there exists no experimental material on relevant heavy-ion reactions, one may make a comparison with experiments on the decomposition of inorganic molecular ions by γ-radiation or energetic electrons[119]. The decomposition reactions in this field are characterized by the so-called G value. This quantity is the number of molecules decomposed, formed or reacted per 100 eV of energy absorbed. Typical G values for the decomposition of compounds like inorganic nitrates or sulfates are $0.1 - 1$ decompositions/100 eV. Applying the lower value to the decomposition of Al$_2$O$_3$ (which would mean oxygen formation) one needs again 7×10^{13} ions/cm^2 of 100 keV for the process. The significance of this brain exercise lies in the fact that both approaches give the same result which at least for the case of the atomic displacement, underestimates the dose necessary for complete decomposition considerably. In practice, for compounds with small G values (< 0.1), one may therefore estimate the order of magnitude of the critical dose without taking electronic processes into consideration. Obviously accurate values must be found in an experiment. Additionally the number of displacements and their recovery depends very critically on temperature. The temperature dependence of the electronic processes, on the other hand, should be much less pronounced. Unfortunately, no data on this problem where heavy-ion bombardment is concerned exists.

If we compare the experimental results mentioned below, we have to take into consideration that for metals the amorphization of the crystal is of major importance, while for covalent compounds the formation and rupture of bonds may take place in the crystalline phase as well as in the amorphous one.

We have already mentioned the case of α-Al$_2$O$_3$ which was studied in detail by Drigo et al.[120] and Naguib et al.[121] with the α-backscattering method. It was shown that after 40 keV Kr$^+$ implantation the damage reaches a saturation level at doses of 10^{16} ions/cm^2. From electron microscope examinations they guessed that the saturation corresponds to the formation of an amorphous layer. 100 keV Pb$^+$ bombardment results in a saturation at $\approx 2 \times 10^{15}$ ions/cm^2. The measured number of displaced Al and O atoms indicates that every lattice atom has at least moved once from its initial position. Nevertheless, it is not clear whether a completely disordered region is formed. This is consistent with the findings of a rather high fraction of Pb$^+$ on substitutional positions along the Al rows even at high doses. Another interesting result arises from the experimental evaluation of the mean values of the displacement energies. They are energy dependent starting with 14 eV for 20 keV implantations and ending up at 96 eV for 100 keV. These facts indicate the role of the re-ordering process and of the electronic interaction during the implantation.

Morhange et al.[122] used ESR and Raman spectroscopy for the study of irradiated diamond. They found a disorder equilibrium at 5×10^{15} N$^+$ ions/cm^2 of 0.55 MeV at RT. At 650°C the equilibrium was reached at 5×10^{17} N$^+$ ions/cm^2. The Raman scattering experiments indicated a saturation dose near 10^{16} ions/cm^2 for 70 keV N$^+$ implantation.

The compound semiconductors attracted the attention of several workers. As_2Se_3 and MoS_2 [123] were bombarded with 100 keV Ar^+ at 35°K and the damage studied with optical absorption spectroscopy and conductivity measurements. It could be shown that above 10^{13} ions/cm² amorphous zones in As_2Se_3 begin to overlap, a process which starts in MoS_2 at 10^{14} ions/cm² and leads to the formation of a continuous amorphous film at 10^{15} ions/cm². This is a rare example of a low-temperature experiment with compounds. Heating to RT causes a considerable annealing and higher temperature leads to recrystallization. Experiments on CdS_2 have been reported in some papers. Hutchby et al.[124 – 126] implanted F^+, Cl^+, Br^+ and I^+ and used optical reflection spectroscopy and ion backscattering for analyzing the damage produced. The saturation dose for the formation of disorder was $10^{16} – 10^{17}$ ions/cm² depending on the ion used. The disorder of a random oriented crystal was definitely not reached for the lower value connected with the heavier ions. Annealing, even for the highest doses, resulted in considerable restoration of the lattice. Walsh[127] also looked into CdS and ZnO damaged with 200 keV Ar^+ ions. With doses of 10^{16} ions/cm² there was still a discernable amount of crystallinity left in both compounds. The dose necessary for amorphization was estimated to be $\approx 2 \times 10^{16}$ ions/cm². The remaining studies are not so much concerned with the formation of amorphous zones or layers as with the dose dependence for the formation of compounds between the target and the implanted ions. In the work of Gruen et al.[128] very high doses of hydrogen were implanted in silicon. Here the host was almost amorphous. Nevertheless, chemical trapping of hydrogen, which indicates the formation of Si-H bonds, occurred. Fig. 27 shows the trapping efficiency as a function of the bombarding dose. At 4×10^{18} H^+ ions/cm² (15 keV) 35 % of the hydrogen formed Si-H bonds. Only at still higher doses, up to 3×10^{19} ions/cm² m the radiation effects lead to a considerable decrease of the Si-H production. This illustrates that the release of hydrogen as H_2 from the target probably counteracts the chemical trapping tendency. It is interesting to note that upon annealing the crystallinity of the Si, including the chemically trapped hydrogen, is recovered. Mohs, Sahm and Wolf[129 – 131] examined the radiation effects of heavy ion bombardment on organometallic and complex compounds. To monitor the radiation decomposition, they made use of the tendency of the implanted ions to react chemically with the host material.

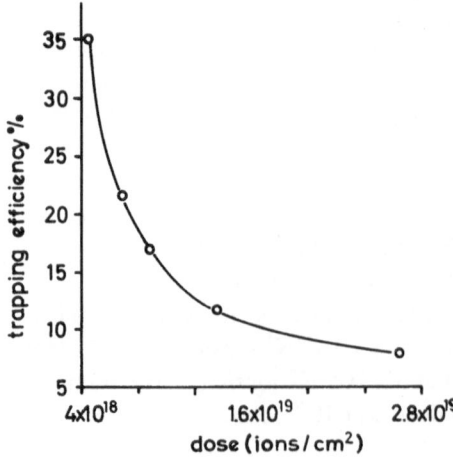

Fig. 27. The trapping efficiency for 15 keV H^+ ions in Si as a function of the dose (after[128])

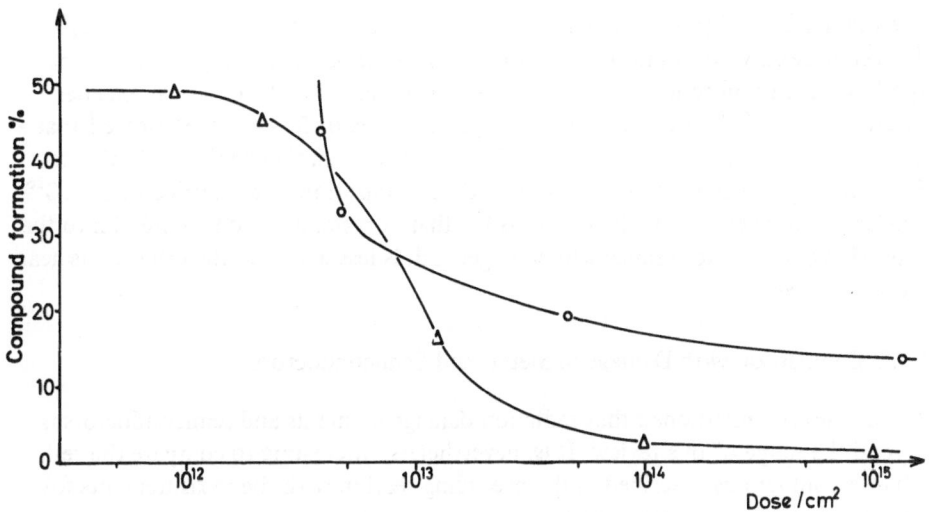

Fig. 28. Formation probability of $^{51}Cr(CO)_6$ for $^{51}Cr^+$ ions in $Mo(CO)_6$ and $[^{57}Co(en)_2Cl_2]NO_3$ for $^{57}Co^+$ in $[Co(en)_2Cl_2]NO_3$ as function of the implantation dose

For the analysis of the products chemical procedures were used. The implanted ions were radioactive in order to be able to distinguish them from the host atoms. An isolated $^{51}Cr^+$ ion implanted in $Mo(CO)_6$ reacts either with the fragments formed during slowing down or with complete $Mo(CO)_6$ molecules through a radiation induced exchange reaction under formation of $^{51}Cr(CO)_6$. The probability for this process is at RT about 50%. Fig. 28 shows how the formation probability develops under increasing dose conditions.

The interpretation of these results is as follows:

As long as we deal with isolated damaged zones in the $Mo(CO)_6$ host the reaction probability remains 50%. At doses $\geq 2 \times 10^{12}$ ions/cm^2 the damaged zones start to overlap and the reaction probability decreases due to decomposition of $^{51}Cr(CO)_6$ formed beforehand. At $\geq 10^{14}$ ions/cm^2 the decomposition is nearly complete and the probability for a restoration of the metal carbonyl from the fragments is very low. The reason for this is probably either a segregation of metallic Mo and gaseous CO or the rupture of the C=O bond under formation of oxygen, carbon, Mo or molybdenum carbide. From the point where overlapping starts one may deduct the size of the interaction zone for a single $^{51}Cr^+$ ion. Assuming a cylindrical zone as deep as the mean projected range one gets a diameter of 80 Å. The dose dependence of the formation probability does not tell us anything definite about the disorder of the $Mo(CO)_6$ lattice. We can only say that crystalline regions are preserved as long as the damaged zones do not overlap. On the other hand, certainly also in amorphous $Mo(CO)_6$ chemical reactions with implanted $^{51}Cr^+$ are possible. Thus we estimate the dose necessary for the amorphization of $Mo(CO)_6$ to be somewhere between 10^{13} and 10^{14} ions/cm^2.

The second curve in Fig. 28 shows the same kind of reaction probability for $^{57}Co^+$ implanted in trans-$(Co(en)_2Cl_2)NO_3$. Here the overlapping of damaged zones seems to start at around the same dose value, but even at 10^{15} ions/cm^2 there remains

49

a reaction probability of $\approx 10\%$ for the formation of *trans*-$[^{57}Co(en)_2Cl_2]NO_3$. The damage necessary to obtain the conditions under which the complex cannot be synthesized any more is much higher than in the case of $Mo(CO)_6$. Mössbauer measurements of the same system[132] at dose values of $3 - 4 \times 10^{15}$ proved that a significant fraction of the $^{57}Co^+$ is in a lattice position similar to the Co in the host. Summarizing, one may say that a complete amorphization of the lattice below 10^{16} ions/cm^2 is not very probable. It is possible that "crystalline" zones extend into the region beyond the mean range which, in general, is less damaged than the layers near the surface (Sect. 2.3).

5.1.2 Comparison with Damage in Metals and Semiconductors

We have already mentioned that radiation damage in metals and semiconductors is beyond the scope of this review. It is, nevertheless, interesting to compare the results for compounds described in the preceding section with the measurements for metals and elemental semiconductors.

It is very difficult to bombard most metals to an extent high enough to achieve the crystalline to amorphous transition. This requires in any case doses $\geq 10^{16}$ ions/cm^2. Very often recrystallization processes take place even at RT. Thus one may get a saturation dose but not a completely amorphous structure. The situation is quite different for the semiconductors Si and Ge which easily suffer a transformation from the crystalline to amorphous states at doses of $10^{14} - 10^{15}$ ions/cm^2. These

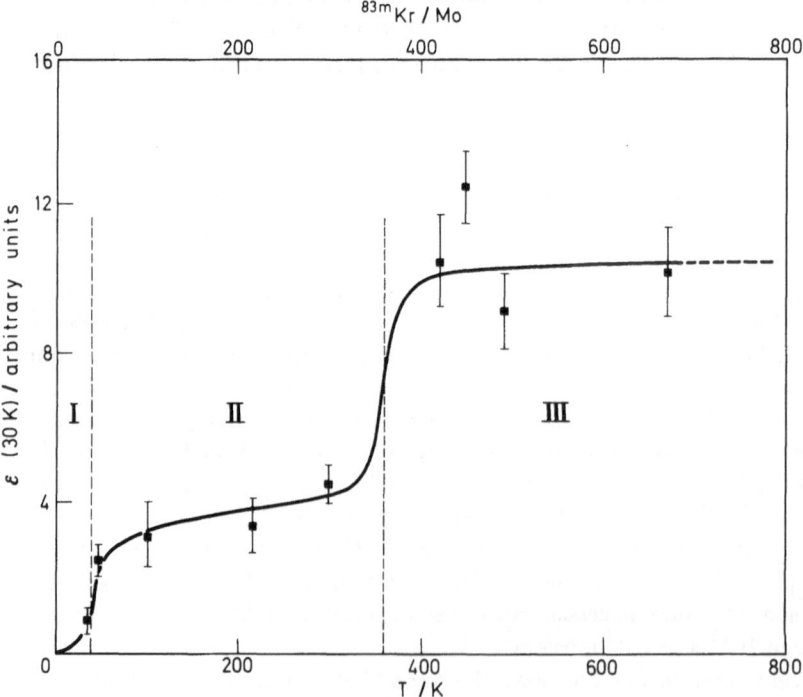

Fig. 29. The recovery of Molybdenum after ^{83}Kr implantation with temperature as measured by Mössbauer spectroscopy[133]

considerations are true at RT or above. In the low temperature region the situation is quite different and a high degree of disorder is caused by rather low doses. In Sect. 2.6.2 we explained the recovery behavior of metals irradiated at liquid He temperature (Fig. 13). A plot of the resistivity against temperature showed that at RT $\simeq 90\%$ of the initial damage had recovered. A similar plot may be obtained by implanting Mössbauer nuclides at different temperatures in metals. As already indicated the Mössbauer nuclide acts to some extent as a probe for the surrounding damage[78].

Table 4. Dose necessary to produce saturation damage in different targets together with the irradiation conditions

Target	Ion	°C	Energy (keV)	Saturation dose (ions/cm^2)	Remarks
Si	He	RT	40–100	$\approx 10^{18}$	
	N	RT		10^{16}	
	P	RT		7×10^{14}	Amorphization
	Sb	RT		$\approx 10^{14}$	
	Bi	RT		5×10^{13}	
Ge	S	RT	40	5×10^{14}	
InSb	Pb	RT	40	$10^{14} - 10^{15}$	Amorphization
GaAs	S	RT	40	5×10^{14}	
Au	Xe	RT	40	$> 10^{16}$	Not completely
W	Kr	RT	40	$> 10^{16}$	disordered
Diamond	N$^+$	RT	70	$\approx 10^{16}$	
NaCl	Xe$^+$	RT	40		
KBr	Xe$^+$	RT	40	$> 2 \times 10^{16}$	Not completely
MgO	Xe$^+$	RT	40		disordered
SiO$_2$	Xe$^+$	RT	40		
Al$_2$O$_3$	Kr$^+$	RT	40	10^{16}	Amorphization
	Pb$^+$	100		2×10^{15}	Not completely disordered
ZnO	Ar$^+$	RT	200	2×10^{16}	Amorphization
CdS$_2$	Ar$^+$	RT	200	2×10^{16}	
	Bi$^+$	RT	25	10^{16}	Not completely disordered
	I$^+$	RT	40	10^{16}	Not completely disordered
	Br$^+$	RT	40	$10^{16} - 5 \times 10^{16}$	Not completely disordered
	F$^+$	RT	40	10^{17}	Amorphous
As$_2$Se$_3$	Ar$^+$	−238	100	10^{13}	Amorphous zones
MoS$_2$	Ar$^+$	−238	100	10^{15}	Amorphous layers
Mo(CO)$_6$	Cr$^+$	RT	60	$10^{13} - 10^{14}$	Estimate
trans-		−100			
[Co(en)$_2$Cl$_2$NO$_3$]	Co$^+$	−RT	60	$> 10^{15}$	Estimate

G. K. Wolf

In Fig. 29 a "recovery" curve of Mo is shown as monitored with the implanted short-lived isotope $^{83}Kr^+$. The measurement has been performed using a Mössbauer spectrometer coupled on-line to an ion accelerator (Sect. 4.2.4).

The curve shows the factor E which is a measure of the recoil-less fraction of the implanted atoms at 30 K. At very low temperatures this fraction is small, indicating a nearly complete disorder around the position of the Kr.

With increasing temperature an ordering process takes place in which clustering of Mo interstitials as well as of krypton atoms and at higher temperatures interactions with vacancies probably take place.

These and numerous other experiments prove that in metals the implanted atoms change their position and their surroundings between liquid He temperature and RT and that a considerable reordering of the lattice takes place even at low temperatures. The low-temperature recovery of ion-bombarded compounds is unfortunately completely unknown. Very few experiments at liquid-N temperatures indicate a strong temperature dependence too. A recovery of the majority of the point defects and centers below RT was found experimentally only for ionic crystals irradiated with electrons[134].

In Table 4 the irradiation conditions for various metals, semiconductors and compounds together with the dose at which the radiation damage tends to saturate are listed. This does not necessarily mean complete amorphization. The values for Si give an impression of the increasing magnitude of the effects if one goes from light to heavy ions. The data are from the studies mentioned above and from[135, 136].

In Table 5 the few data available for the formation and destruction of compounds between the implanted ion and the host are specified. The initial percentage is listed at which the compound is formed under low-dose conditions together with the dose necessary to decompose $\approx 90\%$ of the initial compound.

Table 5. Decomposition of compounds formed during ion implantation. Low-dose formation yield and dose necessary for 90% decomposition

Target	Ion	Energy (keV)	°C	Compound formed	Low-dose yield	Dose for 90% decomposition
Si	H^+	15	RT	Si–H	35% (10^{18} Ions/cm^2)	4×10^{19} Ions/cm^2
$Mo(CO)_6$	^{51}Cr	60	RT	$^{51}Cr(CO)_6$	50% (10^{12} Ions/cm^2)	10^{14} Ions/cm^2
trans-$[Co(en)_2Cl_2 NO_3]$	$^{57}Co^+$	60	RT	$[^{57}Co(en)_2 Cl_2]NO_3$	50–70% (10^{12} Ions/cm^2)	$> 4 \times 10^{15}$ Ions/cm^2
Al_2O_3	D^+	15	RT	AlOOD	Nearly 100% (1×10^{17} Ions/cm^2)	2–3×10^{18} Ions/cm^2

5.1.3 Sputtering of Compounds

In the last section of the chapter on radiation chemistry we shall mention briefly the "radiation" decomposition of surface compounds due to the sputtering effect (Sect. 2.5). There are several reasons why this effect is of interest.

1. During high-dose irradiations of compounds the different elements may be sputtered at different rates. Thus the composition of the surface is changed compared with the one of the bulk of the material.

2. The implantation of reactive ions in an elemental target very often leads to the formation of compounds. If the target erodes during the sputtering process, the surface composition of these compounds is different from the composition below the surface. Thus high-dose oxygen implantation in iron may result in a Fe_2O_3 layer. Sputtering transforms this layer at the surface into Fe_3O_4 [137].

3. Sputtering is a routine method for surface cleaning. The preferential sputtering of the contaminant or the bulk material has naturally some influence on the course of the process.

In general stoichiometry changes on the surface are caused either by preferential sputtering, preferential vaporization from thermal spikes or preferential precipitation of displaced atoms. Preferential sputtering is probably related to the surface binding energy, which one may approximate by the heat of atomization. Some authors propose therefore that the stoichiometry of a target will vary in a way leading to the composition with the lowest heat of atomization [138, 139]. Sputtering induced amorphization of the surface of a compound, on the other hand, should occur above a certain ratio between the crystallization temperature and the melting point (> 0.3) [139]. The alkali halides are sputtered without change of composition [140]. The oxides, on the other hand, may be grouped into different classes. Kelly et al. [141] found that Al_2O_3, MgO, Nb_2O_5, SiO_2, Ta_2O_5, TiO_2, UO_2 and ZrO_2 conform with the theory of collisional sputtering of Sigmund (Sect. 2.5), some with and others without change of stoichiometry due to preferential sputtering of oxygen.

A second group consisting of MoO_3, SnO_2, V_2O_5 and WO_3 have sputtering yield values which are much too high compared with the theory. A thermal contribution to sputtering is probably responsible for this effect.

A third group, MoO_3, Nb_2O_5, TiO_2, V_2O_5 and WO_3, includes the oxides losing preferentially oxygen with or without thermal contribution. The radiation induced change of stoichiometry is sometimes a process during which the oxides remain crystalline ($Fe_2O_3 \rightarrow Fe_3O_4$, $CuO \rightarrow Cu_2O$). The amorphization of the surface without decomposition (Al_2O_3, Ta_2O_5) is also quite common. Finally there are some cases where the initial oxide grows amorphous first and then recrystallizes as new compound. Nb_2O_5, for example, is amorphous after a bombardment with 10^{16} O^+ ions/cm^2 of 35 keV while 10^{17} ions/cm^2 induce a transformation into crystalline NbO [142]. The same is true for rutile-type TiO_2. About 5×10^{15} Kr^+ ions/cm^2 (30 keV) cause amorphization. Up to 8×10^{16} ions/cm^2 the growth of small regions of Ti_2O_3 goes on and beyond 8×10^{16} ions/cm^2 one observes the complete transformation of the surface into polycrystalline Ti_2O_3 [143].

For further information on this subject and on the question of sputtering of surface contaminants the reader should consult the relevant reviews and original papers[138, 139, 141, 144, 145].

5.2 Solid State Chemistry of Implanted Atoms

The number of studies concerning the physical and chemical state of implanted atoms in compounds is very small in contrast to those concerning metals. In this section the most important work on this subject is reviewed. We shall divide the compounds into two classes: the ones with ionic bonds and the ones with covalent bonds. Studies of metals are only mentioned insofar as implantations of reactive non-metals in metals are concerned, a situation leading mostly to compound formation.

Chemical effects in semiconductors and thin films as well as the formation of gaseous reaction products in solids are treated separately. Another field of science producing results of interest for implantation chemistry is the chemistry of hot-atoms resulting from nuclear reactions. Especially in the case of recoil implantation where the energetic hot-atom is produced at the surface of a sample or grain before penetrating into the specimen, one finds a situation similar to ion implantation. The reader should consult the reviews on hot-atom chemistry[94-96, 146-148].

5.2.1 Ionic Compounds

The first "chemical" study of ions implanted in compounds was performed in 1965 by Andersen and Sørensen and is typical for a number of investigations that followed. In all cases the bombarding ions were radioactive and the product analysis was destructive: the sample was dissolved after the implantation and the distribution of the radioactive ions among the different fractions of the analytical procedure was studied. Andersen and Sørensen[149, 150] bombarded K_2CrO_4 and $K_2Cr_2O_7$ with $^{51}Cr^+$, dissolved the targets in aqueous solutions containing Cr^{3+} and CrO_4^{2-} as carriers and separated the solution into two fractions containing the valence states I – III and IV – VI. They found the majority of the ^{51}Cr in the CrO_4^{2-} fraction but a considerable part also occurred as Cr^{3+}. Annealing led to a conversion of the Cr^{3+} into CrO_4^{2-}. Similar experiments were performed with K_2BiF_4 and K_2SO_4. In these targets the presence of $^{51}Cr^{2+}$ was established. Later on the state of implanted Selenium in K_2SeO_4 and Na_2SeO_4 was studied and the existence of Se^0 proven[151].

Another system which has been studied is ^{32}P and ^{35}S implanted in alkali chlorides[152]. Again a destructive analysis of the oxidation states was performed and, for example, S^0 and S^{2-} identified. Modest heating converts the S^{2-} to the S^0 state. A comparison with the same elements originating from nuclear reactions mainly reflects a higher density of defects in the implantation case. Another difference may arise because nuclear recoil atoms very often have a high charge state due to Auger charging while the implanted ion initially has the charge +1 or 0. In addition one very often finds a surprising similarity between results on the behavior of recoil atoms and implanted ions[153 – 155].

The destructive analysis used by the authors mentioned above unfortunately does not allow any safe conclusion about the real oxidation state, the position and the defect structure around the implanted ion in the solid host. The number of papers published using non-destructive techniques is very small and the authors have often not been particularly concerned with the chemical state of the implanted atom.

The Mössbauer effect was used as a non-destructive method of studying the chemistry of ^{153}Sm ions implanted in alkaline-earth fluorides[156, 157]. The most interesting result was the variation in the oxidation state of the ^{153}Eu originating from the ^{153}Sm $\xrightarrow{\beta}$ ^{153}Eu decay which populates the Mössbauer level. While in MgF_2 and CaF_2, with ionic radii of the cations smaller than Eu^{2+}, the Eu was found partly $(10 - 20\%)$ in the +2 state, the SrF_2 and BaF_2 with bigger ionic radii contained only Eu^{3+}. The interpretation of this surprising result is complicated because the Mössbauer spectrum contains no information on the lattice position of Eu, the quadrupole splitting being too small, and the spectrum reflects the combined result of the implantation and the β-decay ^{153}Sm $\xrightarrow{\beta}$ ^{153}Eu ("after-effect").

The authors give the following explanation: Since the implantation temperature was $\simeq 300°$K, a considerable reorientation of the lattice took place during the bombardment. This lead to heavily damaged regions and others with less damage. In the former the thermodynamically most stable +3 state is favorable for the Sm/Eu. In the latter the Sm^{+2} may be stabilized on the lattice positions of the cation. In MgF_2/CaF_2 this requires the presence of anion vacancies as F-centres because of lack of space. The transition Sm $\xrightarrow{\beta}$ Eu initially converts the Sm^{2+} into Eu^{3+}. The charge compensation then requires a fluoride ion on an interstitial position, but since there is a lack of F^- around the Eu^{3+}, the +3 state is reduced to the +2 state, probably with the help of the electrons from the F-centres. In SrF_2/BaF_2 the vacancies or F-centres are not required for the stabilization of Sm^{+2}. Therefore, and because of the larger mobility of interstitials in these compounds, the Eu^{3+} resulting from the β-transition may compensate its charge immediately with an interstitial ion. In addition, oxygen impurities seem to play a role in these systems. Annealing leads to the conversion of the Eu^{2+} to Eu^{3+} and a variation of the chemical shift towards higher electron density, as one would expect for O^{--} taking place of fluoride ions.

In conclusion, it seems that in this case the final state of the implanted atoms at room temperature is the result of a complicated reaction sequence: damage — recovery — β-decay.

With the system $^{57}Co^+$ implanted into FeF_2 the study of the Mössbauer isotope ^{57}Fe in FeF_2 is possible. The iron fluoride molecules develop a long range order below a certain temperature (Néel temperature). The "normal" Mössbauer absorption spectrum changes at this temperature from a quadrupole doublet to a magnetically split complicated spectrum. The emission spectrum arising from implanted $^{57}Co/^{57}Fe$[158] shows this change at a much lower temperature. The lowered Néel temperature indicates that radiation doses of $\approx 10^{15}$ ions/cm^2 disturb the long-range order. On the other hand the short-range order is preserved, the FeF_2 molecules are not destroyed in their majority and the implanted $^{57}Co/^{57}Fe$ comes mainly to rest as $^{57}FeF_2$.

Gruen et al. studied the chemical effects of the implantation of hydrogen and deuterium (15 keV) in sapphire (α-Al_2O_3) with the infrared absorption technique[159].

For the D^+ bombardment at 7.5×10^{16} ions/cm^2 a band assigned to $-$OD was observed. This indicated the formation of Al-O-OD. For the H^+ implantation it was nearly impossible to detect the $-$OH band which was just above the limits of detection. This difference was interpreted as a consequence of the higher damage caused by D^+ leading to an increased reactivity of Al_2O_3. At higher doses the trapping efficiency for D^+ as $-$OD decreased from 1 at 10^{17} ions/cm^2 to 0.07 at 3×10^{18} ions/cm^2. Above 10^{17} ions/cm^2 probably the Al$-$O$-$OD decomposes and D_2 clusters or gas bubbles are formed as indicated by the occurrence of visible blisters. Finally, the authors observed a strong tendency of the irradiated Al_2O_3 to undergo hydrolysis when exposed to water vapors. A similar $-$OH formation was discovered by Roy and Greer for the implantation of $1-5$ MeV protons in TiO_2[160].

The last two examples deal with systems not studied from a chemical point of view but nevertheless of interest for the chemist. The first one is the implantation of Pb in Al_2O_3[161], investigated with regard to the resulting lattice disorder (Sect. 5.1). Lattice location experiments using single crystals showed at low doses ($\sim 10^{14}$ ions/cm^2) $\approx 70-80\%$ of the lead atoms at substitutional positions along Al rows. Even at doses causing considerable disorder $\approx 20-30\%$ of the lead atoms were still in substitutional positions. Since the lattice location was performed by means of the a-backscattering method, the measurements contained no information on the chemical state of the lead ions. But there seems to be no doubt that at least in the substitutional fraction the lead presents itself in the chemical form of an oxide.

The second example is concerned with the implantation of Ag^+ and Au^+ ions into lithia-alumina-silica-glass, in order to induce catalytic surface crystallization[162]. Optical extinction measurements and Rutherford backscattering have shown that the Ag and Au atoms occur at low temperatures as single atoms or very small aggregates. During annealing to 800°C the Ag and Au atoms cluster together to colloids of $20-50$ Å diameter. The Ag colloids are less stable than the Au ones and dissolve into the glass at $200-400$°C. Unfortunately, one does not know whether at least silver forms an oxide either at low temperature or after dissolution. There is no doubt, on the other hand, that the colloids consist of metallic Ag and Au.

5.2.2 Covalent Compounds

In ionic crystals the position of the implanted ion in the lattice, the lattice disorder and the presence of impurities determine the chemical state of the implant. In covalent compounds the formation and rupture of bonds is an additional factor.

Most papers on ion implantations in covalent compounds report destructive methods of analysis. As with ionic compounds, only a few studies exist on non-destructive techniques.

5.2.2.1 Organo-metallic Compounds

One of the first investigations was the implantation of Cu^+ in α- and β-phthalocyanine. It was found that the percentage of Cu bound to the host was higher in the α-form[163]. Later on Yoshihara et al.[164] bombarded metal-free and copper phthalocyanine with radioactive In^+, $InCl^+$ and $InCl_3^+$. After dissolution of the target the complexed and uncomplexed fractions were separated. From the fact that $\approx 5 - 30\%$ of the radioactive indium were measured in the complexed fraction the authors concluded that indium-phthalocyanine was the main reaction product.

Another attempt to synthesize organo-metallic compounds was made by Sahm and Wolf[165, 166]. They found that $\sim 50\%$ of $^{51}Cr^+$ implanted in $Cr(CO)_6$ or $Mo(CO)_6$ was converted to "organic" ^{51}Cr, being volatile and extractable into organic solvents.

The conclusion was that $^{51}Cr(CO)_6$ had been formed during a reaction between the implanted ^{51}Cr and the target molecules or their fragments arising from the radiation decomposition.

Thin-layer chromatography was used to prove the identity of $Cr(CO)_6$ and the radioactive ^{51}Cr-compound. A similar reaction occurred between ^{51}Cr and $Re_2(CO)_{10}$ but the "organic" yield was much lower ($\approx 17\%$). The position of the $^{51}Cr^+$ between $Cr(CO)_6$ and $Re_2(CO)_{10}$ in the thin-layer chromatogram was interpreted as an indication for the possible formation of a mixed Cr-Re-carbonyl. Different radioactive ions were then implanted in various carbonyl compounds and in nearly all cases an "organic" radioactive fraction between 4 and 20% was isolated. Table 6 contains all systems studied.

Because of the low number of molecules formed ($10^{13} - 10^{14}$) during an irradiation a direct identification of the structure of the reaction products was impossible. The behavior of the compounds during thin-layer chromatography showed a certain probability that at least for reactions between homologues ($Co+Ir(CO)_{12}$, $Fe+Ru_3(CO)_{10}$) the corresponding mixed carbonyls ($CoIr_3(CO)_{12}$) had been formed. In all other cases one can only guess the identity of the "organic fraction". The most interesting outcome was the "organic fraction" originating from the reaction be-

Table 6. Formation of organo-metallic compounds by ion implantation

Implanted ion	Target compound	% Organic yield Sublimation and/or solvent extraction
$^{51}Cr^+$	$Re_2(CO)_{10}$	14,5
$^{56}Mn^+$	$Re_2(CO)_{10}$	7
$^{57}Co^+$	$Re_2(CO)_{10}$	6,4
$^{57}Co^+$	$Ru_3(CO)_{10}$	6,6
$^{57}Co^+$	$Rh_4(CO)_{12}$	4,5
$^{57}Co^+$	$Ir_4(CO)_{12}$	3,8
$^{59}Fe^+$	$Re_2(CO)_{10}$	0,6
$^{59}Fe^+$	$Ru_3(CO)_{12}$	17,5
$^{59}Fe^+$	$Rh_4(CO)_{12}$	6,1
$^{59}Fe^+$	$Ir_4(CO)_{12}$	10,0
$^{64}Cu^+$	$Mn_2(CO)_{10}$	3,0
$^{64}Cu^+$	$Ru_3(CO)_{12}$	8,6
$^{64}Cu^+$	$Rh_4(CO)_{12}$	7,0
$^{64}Cu^+$	$Ir_4(CO)_{12}$	7,5

tween $^{64}Cu^+$ and carbonyl compounds. Until now no volatile and stable copper compound is known which may result from these conditions.

Implantations in carbonyl compounds were also studied by Wiles and co-workers[167, 168]. They implanted $^{56}Mn^+$ in $Cr(CO)_6$ and analyzed the organic fraction. Rather low but significant yields ($< 1\%$) of "organic" ^{56}Mn were found and the possibility for the formation of a mixed carbonyl discussed. In the work of Sahm and Wolf mentioned above the organic yields were much higher but the system $^{56}Mn+Cr(CO)_6$ was not studied. The discrepancy might arise from the very low ion energy in the work of Wiles. The penetration of a 5 keV $^{56}Mn^+$ into the target is so small that surface reactions should play a major part. Another study of Kanellako-pulos and Wiles[169] on $20 - 80$ keV $^{59}Fe^+$ in $(CpFe(CO)_2)_2$ and $CpFe(CO)_2$ gave higher "organic yields". The main product ($\approx 5\%$) was $(Cp^{59}Fe(CO)_2)_2$; $Fe(CO)_5$ and $FeCp_2$ were, to a lesser extent, also released. The authors assume that a hot "exchange" reaction took place during the formation of the $(Cp^{59}Fe(CO)_2)_2$ and a reaction of $^{59}Fe^+$ with free ligands which lead to the synthesis of $Fe(CO)_5$.

The last organo-metallic compound to report on is Nd-hexafluoroacetylaceto-nate ($Nd(HFA)_3$). The effect of the implantation of rare-earth ions of 40 keV was compared with the one from recoiling atoms produced by spallation reactions with 660 MeV protons[170]. The chemical form of the radioactive products was determined by dissolving the target and applying distribution chromatography or studying the volatility during vacuum sublimation. The result showed a considerable percentage of the radioactive lanthanide ions in the parent form substituting the Nd of the target. The chromatography gave an organic yield twice as high as the sublimation probably because of stabilizing reactions with the solvent. The comparison with the reactions of recoil atoms from nuclear reactions yielded an astonishing degree of agreement.

Organic Compounds: Very little work has been done on the interaction of ion beams with solid organic compounds. Studies to be mentioned are the ones on reactions between J^+ and J_2^+ ($5 - 3000$ eV) with solid alkanes[171] and ^{14}C ($2 - 5000$ eV) with benzene[172]. In both cases considerable fractions of labelled organic iodides and hydrocarbons (^{14}C labelled benzene, toluene, etc.) respectively were found.

5.2.2.2 Complex Compounds

Some studies dealt with the compound $(Co\,(ethylendiamin)_2Cl_2)NO_3$. The advantage of this compound is the occurrence of a *cis-* and *trans-*isomer. Thus the isomerization might act as a probe for the probability of smaller changes in the lattice structure without permanent destruction of the complex bonds. Also there exists information on this system from studies of nuclear recoil atoms. The early papers of Andersen, Langvad, Sørensen[163] and Wolf and Fritsch[173] report on results obtained by destructive analysis of the products of the reaction $^{57}Co^+ + (Co(en)_2Cl_2)NO_3$. The configuration of the host is mostly preserved in agreement with the findings of the recoil atom experiments[174, 175, 177]. At rather low doses around $10^{13}\,Co^+/cm^2$ about $20 - 50\%$ of the $^{57}Co^+$ atoms were found in the same chemical form as the target, the percentage being higher for the *trans-*$(Co(en)_2Cl_2)NO_3$ host compared with the *cis-*compound. The fraction undergoing isomerization was low in the *trans-*host but rather high ($\approx 30\%$) in the *cis*-host. This resembled the lower stability of *cis*-(Co-(en)$_2$Cl$_2$)NO$_3$ in comparison with the *trans*-configuration. Mohs[176] extended the measurements to higher bombardment doses ($\approx 10^{15}$ ions/cm^2). The ^{57}Co$^+$ fraction incorporated in the host decreased with increasing doses but was still considerable. At higher dose values the *cis*-configuration was found in the *trans*-compound too.

More information on this system came from non-destructive Mössbauer measurements[132, 178]. In *cis*- and *trans*-(Co(en)$_2$Cl$_2$)NO$_3$ about 10^{15} ^{57}Co$^+$/cm^2 were implanted and the targets used as Mössbauer sources. The spectra were compared with emission spectra of the same compounds labelled with ^{57}Co$^+$ via chemical synthesis. One must keep in mind that the Mössbauer spectrum contains no direct information on the ^{57}Co in the matrix but only on the actual Mössbauer nuclide ^{57}Fe arising from the ^{57}Co $\rightarrow$ ^{57}Fe β-decay. Because the corresponding compounds *cis*- and *trans*-(Fe(en)$_2$Cl$_2$)NO$_3$ were not accessible by synthesis, a comparison with the absorption spectra was not possible. The interpretation followed the line:

a) the spectra contained a number of doublets indicating a few distinct reaction products,

b) the emission spectra of the labelled complexes as well as the implanted ones showed ^{57}Fe as Fe^{II} and Fe^{III} in the nearly undisturbed ligand cage formed by two ethylene diamine and two chlorine having a *trans*-configuration and as Fe^{III} low-spin complex. There was no indication of ^{57}Fe in the *cis*-configuration in the implanted targets neither in a *trans*- nor in a *cis*-host.

c) the emission spectrum of the implanted complexes contained a dominating additional doublet probably arising from Fe^{III} in a distorted or completely rearranged environment.

If we compare these results with the ones obtained by destructive analysis, taking into account the difference between the two systems, we find a satisfactory agreement: the order of magnitude of the ^{57}Co/^{57}Fe found in the *trans*-configuration is reproduced in the spectrum. The compound responsible for the additional doublet is probably the source of ionic cobalt found after dissolution. The only serious discrepancy is the complete absence of ^{57}Fe in the "*cis*-fraction" measured spectroscopically in contrast with the dissolved *cis*-(Co(en)$_2$Cl$_2$)NO$_3$ which contained a considerable amount of ^{57}Co.

^{57}Co/^{57}Fe implanted into K$_4$(Fe(CN)$_6$) and K$_3$(Fe(CN)$_6$) was another combination under investigation by means of non-destructive Mössbauer spectroscopy[179, 180]. These compounds had the advantage of being well known. Thus the absorption spectra and the emission spectra were available for comparison with the spectra from the implanted samples.

Fig. 30 shows the spectrum of ^{57}Co/^{57}Fe implanted in K$_4$(Fe(CN)$_6$) (a) and in K$_3$(Fe(CN)$_6$) (b). Again the spectrum contains information on the existence of a few well-defined species. A total destruction of the lattice and a major breaking of the C=N bonds did certainly not take place though the implantation dose of $\approx 10^{15}$ ions/cm^2 might have damaged the target seriously. The species A, B and C for the K$_4$(Fe(CN)$_6$) target correspond with ^{57}Fe in the position of Fe^{II} in the undisturbed inner sphere (A), a Fe^{II} at the position of a K$^+$(C) and a Fe^{III} bound in the inner sphere to less than 6 CN$^-$ ligands (B). The less stable K$_3$(Fe(CN)$_6$) contains only the species B and C. The mother configuration, a ^{57}Fe in the undisturbed inner sphere, is not preserved.

A comparison with the emission spectra of the labelled compounds[181, 182] shows that in this case the mother configuration in $K_4(Fe(CN)_6)$ is the major product although it accounts only for $\approx 10\%$ of the ^{57}Fe in the implanted sample. For $K_3(Fe(CN)_6)$ also the spectra of the labelled complex show only $\approx 20\%$ of the Mössbauer isotopes to have the configuration of the host. The differences are certainly caused by the additional damage coming from the implantation; on the other

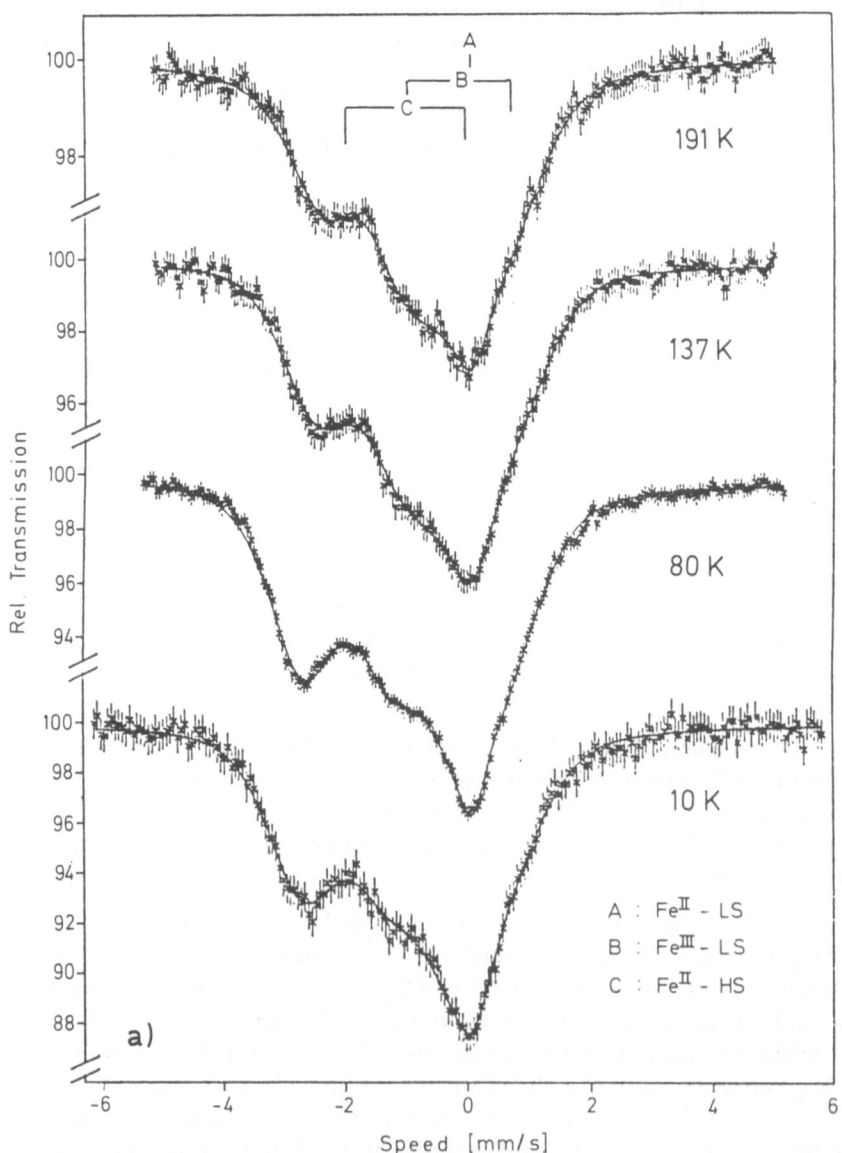

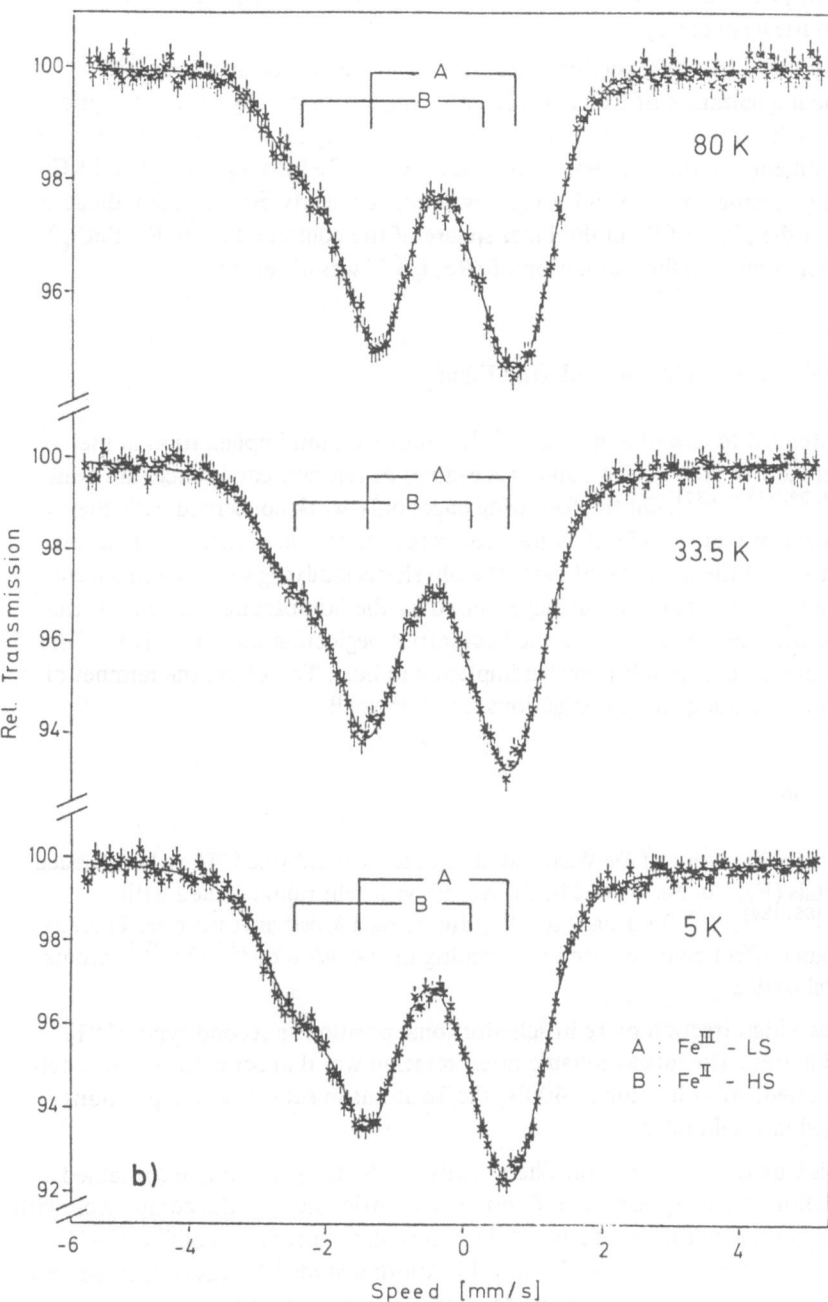

Fig. 30. Mössbauer spectrum of $K_4(Fe(CN)_6$ (a) and $K_3(Fe(CN)_6)$ (b) bombarded with $^{57}Co^+$ ions[180)

hand, the majority of the complex bonds were not destroyed during the bombardment in an irreversible way.

Finally the work of Rössler on hexahalo complexes will be mentioned[183, 184]. He studied the implantation of Re^+ into K_2SnCl_6 single crystals. By means of optical spectroscopy he could show that even at liquid He cooling a Re-Cl bond formation took place. Recoil implantation experiments of radioactive Re into $K_2(SnCl_6)$ and KCl followed by destructive analysis brought evidence for nearly 50 % of the radioactive Re being on the place of Sn in the inner sphere of the complex for the $K_2(SnCl_6)$ host. In KCl no measurable formation of $(ReCl_6)^{2-}$ was observed.

5.2.3 Metals, Semiconductors and Thin Films

It is not intended to describe in detail all the studies on ion implantation in metals or semiconductors. Here reviews and proceedings of relevant conferences are available[1, 2, 19, 34, 185 − 187]. On the following pages only work concerned with the bombardment of metals and semiconductors with reactive non-metals will appear. Unfortunately, in the majority of cases the physicists studying such systems were only interested in the radiation damage caused by the bombarding particles or the position of the implanted atoms in the host lattice neglecting the possibility of the formation of chemical bonds between implant and host. Therefore, the number of papers containing adequate investigations is rather small.

5.2.3.1 Metals

Here especially the work of De Waard et al. should be mentioned. They bombarded several metals (Fe, Co, Ni, V, W, Pb, Cr, Au, P) with tellurium labelled with $^{129m}Te^{78, 188, 189}$. ^{129m}Te decays to ^{129}I, this being a Mössbauer isotope. Therefore the Mössbauer effect could be used for locating the position of $^{129m}Te/^{129}I$ atoms in the metal lattice.

Besides a high fraction of Te in substitutional positions a second type of "Te" was found in iron. The most probable interpretation was that formation of iron tellurides occurred. After heating to 400°C the Te atoms in substitutional positions also changed into tellurides.

The same is true for ^{129}Te in gold. The initially implanted gold sample contained a certain fraction of $AuTe_2$ as well as Te on regular lattice sites, as the comparison with a chemically prepared source of $Au^{129m}Te_2$ showed. On heating to 600°C the Te presented itself completely as $AuTe_2$ in gold. Another study[190] was concerned with ^{129m}Te in α- and β-tin. Here also more than one lattice position was recorded but the question of formation of a compound remains open.

Several authors worked on the problem of the oxidation state of rare earth ions implanted in iron. Cohen, Beyer and Deutch[191] bombarded iron with 50 keV ^{151}Gd ions which decay by β-transition to ^{151}Eu. Mössbauer measurements were used for the analysis of the chemical state of the rare earth ions. The result showed all Eu atoms in the +3 oxidation state but at two different sites. About 60% occupied a substitutional position whilst the rest was present in a form which the authors do not identify as interstitials but as internally oxidized Eu^{3+}. Annealing converts the substitutional Eu atoms into internally oxidized ones.

Another group[192] performed channeling and PAC studies on the annealing of ^{172}Yb in Fe and came to a similar interpretation of their results.

In this context the source of the oxygen is unclear. It might come from impurities or from recoil implantation of the surface oxygen caused by the knock-on from the bombarding Gd ions. Other authors were doubtful about this interpretation[193 – 195]. PAC measurements of 400 keV ^{152}Eu ions in Cu did not indicate any oxidation of the Eu on annealing.

400 keV ^{174}Yb implants were performed with subsequent implantation of ^{18}O in the layers near the surface. An oxygen profile using the nuclear reaction technique (Sect. 4.3) did not show any enrichment of oxygen in the implanted region. Thus the problem of internal oxidation of implanted rare earth ions is still unsolved. A possible explanation of the discrepancies mentioned might be the lower implantation energy used by Cohen et al. Thus the penetration depth of the implanted ions might be just the same as the one of the recoil implanted oxygen, while the higher energy used in the other experiments resulted in an end position of the rare earth ions far beyond the range of recoiling surface oxygen.

A lot of work has been done on the physical consequences of the interaction of hydrogen or deuterium ions with surfaces[196 – 199]. These experiments are important because of the material problems in fusion reactors. The release of hydrogen as well as the trapping may serve as tool for the investigation of damage sites. Chemists will immediately realize that in most materials the formation of hydrides is an important process which certainly plays a role in the studies mentioned. But until now this aspect has received only minor attention and has been roughly sketched with the paraphrase "chemical trapping"[200 – 202]. We shall touch on this question again in the following section.

5.2.3.2 Thin Films

Ion implantation is a method very often used for changing the electrical properties of thin films. Sometimes the bombarding ions may react with the film material. In contrast to the cases mentioned in the last section, thin films change entirely because their thickness has the same order of magnitude as a typical range of light ions (≈ 1000Å). Pavlov et al. studied the phase composition of aluminium films bombarded with As^+, P^+, N^+, C^+, B^+ ions and of iron with C^+ ions[203]. In comparing the results of electron diffraction patterns with the ones of known compounds they concluded that crystalline AlN and Al_4C_3 had been formed. The AlP and AlAs were amorphous at low temperatures but became crystalline upon heating. C^+ in α-Fe reacted to ϵ-Fe_3C, heating to 400°C resulted in a transformation to cementite and α-Fe.

Tantalum is another important material in thin film technology. Bombardment with light ions leads to an increase in resistivity of the thin film. Wilson, Goh and

Stephens[204 – 206] investigated the ion-induced phase changes in evaporated films of 500 Å thickness using electron microscopy, X-ray and electron diffraction. Originally the films consist of a mixture of tantalum, tantalum oxide and oxygen. Argon bombardment leads to a change in the phase structure only when the film looses enough material due to sputtering. Then TaO_2 precipitates are formed possibly by knock-on implantation of oxygen from the surface. Oxygen implantation at low doses creates islands of crystalline *bcc* tantalum and TaO_2 precipitates at higher doses. Nitrogen bombardment finally results in the formation of Ta_4N_5. This is especially interesting because the highest "normal" nitride is Ta_3N_5 which is very difficult to convert with usual chemical techniques into Ta_4N_5.

5.2.3.3 Semiconductors

As in the case of thin-films ion implantation is also used for changing the properties of semiconductors. The bombardment with reactive light ions in this case also often leads to the formation of chemical compounds. The formation of carbides, nitrides, oxides, and hydrides in Si and Ge in particular have been demonstrated[207 – 209]. Borders and Beezhold[85], for example, studied the infrared transmission properties of Si single crystals implanted with high doses of C^+, N^+ and O^+. The spectra showed clearly the formation of crystalline SiC after annealing at 850°C and of Si_3N_4 and SiO_2 after annealing at 1000°C. At lower temperatures one observes broad featureless absorption bands. This does not mean that the formation of the compound takes place entirely at that high temperature but that the compounds are amorphous at lower temperature. Amorphous SiO_2 or Si_3N_4 formation was investigated by a Russian group[210, 211], too. They observed SiC which remained amorphous up to an implantation temperature of 500°C but which changed between 600 – 850°C into β-Si_3N_4. N^+ bombardment above 850°C resulted in polycrystalline β-Si_3N_4. Gruen et al. studied the state of hydrogen and deuterium in silicon, germanium and their compounds[212, 213, 215]. They monitored the formation of Si-H, Si-D, Ge-H, Ge-D bonds and the lattice damage by infrared spectroscopy and laser-induced Raman Scattering. It could be shown that at low doses ($< 10^{16}$ ions/cm^2) mainly physical trapping of the H and D atoms and the implanted ions. The trapping efficiency (no. trapped atoms/no. host atoms) was as high as 1.2 – 2 for Si-H and Si-D, respectively, and 0.8 for Ge-H and Ge-D. One concluded the following: Chemical trapping happens only at doses leading to the amorphization of the lattice. The deuteron trapping is more efficient because deuterons damage the lattice to a higher degree. If one heats the hosts to 630° and 460°C for Si and Ge, respectively, the damage anneals and the host-hydrogen bonds are integrated in the resulting crystalline layers[216].

CdS and GaAs are compound semiconductors studied extensively with respect to lattice disorder and changes of the electrical properties produced during ion bombardment.

There is no doubt that the implantation of P^+ and S^+ into GaAs[214] or of the halides F^+, Cl^+, Br^+, I^+ into CdS[124, 126] leads partly to the formation of chemical bonds between host and implant. But unfortunately this question was never the subject of investigation and remains open.

5.2.4 The Formation of Gaseous Products in Solids

The possibility of the formation of gaseous reaction products in ion-bombarded solids exists either in the case of two different implanted elements reacting with each other or the implanted ions reacting with constituents of the target. There are two fields where such reactions play a major role. The first one is the fusion research where one expects hydrogen to react with the wall constituents of a fusion reactor, the second one is cosmochemistry where one knows that solar wind reacts at the surface of the moon or planets. In both cases the ion implantation technique is able to simulate the real processes.

5.2.4.1 Reactions of Energetic Hydrogen in Solids

In the preceding section we explained the reactions of energetic hydrogen atoms with metals and semiconductors resulting in the formation of hydrides (chemical trapping). In addition volatile products, released from the solid phase under certain conditions, may also occur. Thermal desorption as well as sputtering causes the release process. This leads to an increased erosion of the wall of a fusion reactor by so-called chemical sputtering[196]. A number of experiments have been performed partly using gas discharges partly ion beams. In the majority of the gas-discharge experiments only the erosion rates of the proposed wall materials, mostly C, SiC, B_4C, BN, have been studied[217, 218].

In some experiments a mass spectrometer coupled to an ion source was used for the identification of the released molecules[219]. H_2 discharges and carbon-rich materials generated methane derivatives CH_3^+, CH_4^+, CH_5^+, ... Hydrogen reactions with glass, silica and alumina[220] lead to H_2O and CO formation. In all cases reactions occurred not only with the bulk constituents but also with surface contaminations. The distinction is not difficult because in the case of a pure surface reaction the product yield decays exponentially with time as the surface concentration is depleted.

The experiments with energetic hydrogen or deuterium beams are much easier to interpret because additional reactions in the gas phase as they occur in gas discharges can be excluded. In a number of laboratories the target erosion rate due to methane production has been measured as a function of the substrate temperature[221, 222]. Generally a maximum was observed around 600°C (for carbon), the erosion yield being $3 - 8 \times 10^{-2}$ atoms per incident atom. For SiC and B_4C the erosion yield is much lower than for carbon. The results of all studies suggest the following explanations[223] for the methane formation:
the methane formation does not take place during the initial collision of the incident ion with the target. The ions rather slow down first and diffuse back to the surface where the hydrocarbon is formed.

The reaction mechanism proposed in[224] consists of a stepwise interaction of adsorbed hydrogen atoms with carbon starting with CH_{ads} and ending up with CH_4 gas.

The yield maximum as a function of temperature is probably due to two effects. At low temperatures the reaction rate of the chemical reaction increases with temperature. At high temperatures the recombination of H_{ads} to H_2 with subsequent desorption leads to a lowering of the surface concentration of hydrogen and to a lower methane production rate.

5.2.4.2 Simulated Solar-wind Implantation

The solar wind contains various elements. The most abundant one is hydrogen. The quantities of carbon, oxygen and nitrogen present are $10^{-3} - 10^{-4}$ times less. Experiments with lunar fines showed hydrogen concentrations up to 1 cm^3/g and a carbon or nitrogen content in the region of 100 ppm. The dependency of the content of these elements on the grain size showed that the concentration was much higher at the surface of the material than in the bulk. This enrichment at the surface is attributed to the solar-wind implantation[225, 226].

The different components of the solar-wind may undergo complicated chemical reactions. Nitrogen is present mainly in form of nitride or ammonium salt, the carbon as carbide or hydrocarbons. On heating material from the moon one was able to record H_2, CO, CO_2, CH_4, C_2H_6, C_3H_8, C_2H_4 and C_3H_6 using mass spectrometry and gas chromatography[227].

Ion implantation seems to be very well suited to simulate the reactions of the solar-wind. The energy necessary is $\approx$ 1 keV/nucleon (for $^{12}C \approx$ 12 keV) and easily obtainable with low-energy accelerators. The intensity of the ions in the solar wind is many orders of magnitude lower than in an ion beam. Thus, the conditions of many years of solar-wind impact are achieved in seconds to hours.

One of the first experiments was performed 1972 by Pillinger et al.[228]. They implanted $^{13}C^+$ and D^+ in different materials including lunar fines. By treatment with acids they got $^{12}CD_4$, $^{13}CD_4$ and $^{13}CH_4$ indicating the formation of methane in reactions between the two implants as well as between the single implanted elements and constituents of the moon material.

In another study different minerals which are known to occur on the moon were bombarded with a mixture of $^{13}C^+$, D^+, and $^{15}N^{+}$[229]. It was shown that molecules as light hydrocarbons and HCN as well as carbides had been formed. During heating or sputtering of the target the volatile components were released. The analysis was done using acid hydrolysis combined with mass spectroscopy.

Chang et al.[230] used terrestrial basalt and lunar fines for their simulation experiment with $^{13}C^+$ and $^{15}N^+$. Vacuum pyrolysis of the irradiated targets led predominantly to the release of $^{13}C^+$ as CO and CO_2 and ^{15}N as N_2. Chang and Lennon finally[231] worked on the trapping efficiency of lunar fines for $^{13}C^+$ and ^{15}N. At doses below 2×10^{16} ions/cm^2 about 70 % of both elements are retained in the target, while above 2×10^{17} ions/cm^2 the trapping efficiency is still at the initial level indicating a special mechanism for the trapping of carbon ions. There are many other investigations on the simulation of the behavior of rare gases from the solar-wind in lunar material which we shall not treat in detail in this review[226, 232, 233].

6 Surface Chemistry of Materials Irradiated with Ions

Ion bombardment may lead to changes in the physical and chemical properties of the surface layers of elements and compounds, and, therefore, be of present and future technological interest. The effects are based on surface damage, the formation of amorphous layers or surface alloys without change of the properties of the bulk materials. In this chapter experiments on such surface variations will be reviewed insofar as they are interesting from a chemical point of view.

6.1 Corrosion and Oxidation

The majority of studies on surface chemistry of ion-bombarded samples are concerned with the oxidation and corrosion of materials. One part of the experiments covers the corrosion and oxidation in gaseous atmosphere such as air or oxygen at normal or high temperatures. The other, smaller, part deals with aqueous corrosion, in particular with the dissolution of metals and the formation of passivating layers in aqueous solutions. The interest in this subject found its expression in two conferences in 1975[234] and in 1978[235].

6.1.1 Oxidation and Corrosion in Gaseous Atmosphere

The investigation of the oxidation of surfaces is either interesting from the standpoint of basic science in order to yield information about the mechanism of adsorption of oxygen and the subsequent formation of the oxide or from the standpoint of applied science for a better understanding of the corrosion of materials. In the latter case oxidation may lead to the unwanted growth of oxides in the form of rust or to the wanted growth of passivating layers preventing further corrosion. There is no doubt that ion implantation has some influence on the oxidation process.

In one case the surface is cleaned by the ion beam, in the other a change in its structure takes place. Both events may increase the surface reactivity. In addition, the implantation of appropriate elements produces surface alloys and varies the corrosion properties of the basic material without influencing its other characteristics. A number of authors investigated the oxidation process without the direct aspect of corrosion.

Goode[236] studied the oxidation of polycrystalline nickel bombarded with He, Si, Ne, Ca, Ti, Ni, Co, Xe, Ce, and Bi. The oxides were grown in oxygen at 630°C and the nuclear reaction $^{16}O(d, p)$ ^{17}O was used for the determination of the amount of oxygen take-up. The specimens bombarded with self-ions were annealed at different temperature. The atomic size of the implanted atoms and the annealing of defects were found to have a marked influence on the growth rate of the oxide. The oxidation rate of the implanted region compared with the non-implanted one increased with annealing temperature having a distinct maximum around 400°C. It was concluded that three effects were responsible for the observed behavior:

1. The increasing distortion of the nickel lattice due to the effect of the increasing atomic size

2. the annealing of the defects arising from the bombardment
3. vacancy-assisted cation migration occurring only over a limited temperature range.

One of the early studies was the work of Trillat and Hayman[237] who showed that the oxidation rate of freshly polished uranium was reduced by argon implantation.

A similar passivation effect was reported by Mezey et al.[238] for silicon implanted with $10^{15} - 10^{16}$ Ge^+ and Si^+/cm^2. The tendency for thermal oxidation at $900°C$ was considerably smaller for the implanted than for the untreated silicon. Nitrogen implantation also resulted in passivation of the silicon surface[239]. Other authors found, in contrast, an enhanced oxidation of ion-bombarded silicon. Nomura and Hirose[240] reported a thickening of the oxide layers of Si after implantation of Sb^+, P^+, Sn^+ and Ar^+ with doses $> 5 \times 10^{14}$ ions/cm^2. Williams and Grant[241, 242] bombarded Si with Ga, Cs, Pb, Bi and Xe and measured an oxidation enhancement for Cs and an inhibiting effect for Ga. The transport phenomena involved in the growth of the oxide layers may explain some of the differences. While Pb^+ implanted in Si moves with the interface Si/SiO_2 during oxidation and Bi^+ is found partly in Si and partly in SiO_2, Xe^+ remains completely stationary in the region near the surface.

Transport phenomena during anodization were also studied by Mackintosh and Brown[243]. In Al_2O_3 the halogens and alkali metals moved into the specimen, the former ones to greater, the latter ones to lesser depth than implanted rare gases. Ag, Ba, Ca, Co, Cu, Fe, Ga, Hg, In, Mn, Ni, Sr, Tl, Sb and V moved outwards. In elemental Al the alkali and alkaline earth metals moved towards the solid-electrolyte interface. The remaining metals moved partly with the advancing oxide front outwards into the growing oxide. Fowler et al.[244] implanted Bi, Sn, Pb, Tl, Ce, Kr, Ag, Cr, Cu and Au to a depth of 400 Å in Al. During anodization Bi, Sn, Pb and Tl led to an increased oxygen uptake, Kr showed no effect and Au, Cu, Cr and Hg reduced the oxidation. An analysis using the He^+ backscattering method showed that the oxidation rate starts to decrease as soon as the oxide front reaches the implanted atoms.

The remaining studies on oxidation of metals have been performed mainly with the aim of looking into the corrosion properties of materials. Thus, in 1971 it was found that the implantation of boron generated passivation in copper[245]. The oxidation rate of zircaloy-4 in oxygenated water at $300°C$ was suppressed by ion bombardment[246]. Implantation of reactive ions such as O^+ and chemically inactive ones such as Ar^+ and Xe^+ yielded the same retardation of the oxidation, suggesting that the lattice damage was the main reason for this effect.

Naturally iron and steel are the materials of particular interest in corrosion research. To get an idea of the action of irradiation-induced defects on the oxide layer of iron, Ashworth et al.[247] made measurements using an electrometric reduction technique which gives the thickness of the oxide layer as well as its composition. The results showed that the process of ion implantation with inert gas ions thickens the air-formed oxide film on iron. It is worth noting that this thickening does not so much affect the outer film on iron, consisting of γ-Fe_2O_3, but rather the inner one consisting of Fe_3O_4. This result leads to the suggestion that the variation of the oxide film due to the defects produced during ion implantation might be the reason for the passivation reported in the majority of the other studies mentioned above.

On the other hand, in a number of cases the question of the formation of alloys between the implant and the host or of the migration of oxygen under the influence of the implanted impurity is of importance.

Antill et al.[248, 249], for example, investigated the long-term oxidation behavior of an austenitic steel after implantation of yttrium and cerium. Yttrium was implanted in 20/25 Nb steels and the weight gain measured as a function of the exposure time to CO_2 at 700°C. The implanted sample showed a considerably lower weight gain than an untreated one. The effect was as big as the one achieved by alloying the steel in a conventional manner with 0.1 − 0.4 % yttrium. Implantation of Kr^+ and Nb^+ did not bring any benefit. The implanted yttrium remained effective even for oxide thicknesses much greater than the implanted depth. A depth analysis using secondary-ion mass spectrometry showed an accumulation of yttrium at the metal-oxide interface.

Implantations of yttrium and cerium in 15 % Cr/4 % Al steel and aluminized coatings on nickel-based alloys did not improve the high-temperature oxidation resistance even though conventional yttrium alloy addition had an effect. The differences for the various substrates are attributed to different mechanisms of oxidation of the materials. The austenitic steel forms a protective oxide film and the oxidation proceeds by cation diffusion. Thus, the yttrium is able to remain in a position at the oxide/metal interface. The other materials exhibit oxides based on aluminum. In their growth anion diffusion is involved which means an oxide formation directly at the oxide/metal interface. The implanted metals may, therefore, be incorporated into the oxide and lost by oxide spalling.

The oxidation of titanium and zirconium was examined extensively by Dearnaley et al.[250, 251]. They implanted a variety of up to 30 different elements from all over the periodic table in the two metals. Some of them inhibited the oxidation at 400°C, others did not. Radiation damage, as caused by argon implantation, plays an insignificant part for Ti and Zr. Apart from this, both metals showed striking differences as far as the effects from different implants are concerned. For Ti ions with large ionic radii and large heat of oxide formation (= low electronegativity) caused the highest oxidation inhibition (Ca, Ce, Eu, Cs) and vice versa (Bi, Ni, Sb, Ru). In Zr no such correlation could be found, only the ionic size of the implanted ions seemed to play a role. The following explanation was suggested:

In Ti and Zr the oxidation proceeds by oxygen transport into the metal. In Ti the species which form the perovskite structure ($SrTiO_3$) are very effective in blocking the path for the oxygen diffusion and in inhibiting the thermal oxidation. In Zr it seems that only the stress caused by the implanted ions is responsible. It leads to cracks and ruptures of the oxide along which the oxygen can penetrate and attack the metal. Therefore, the transition metals (Fe, Ni, Cr) having ionic radii similar to Zr, are the best inhibitors. This interpretation is supported by the fact that in Zr the inhibition has a maximum at a dose value of 5×10^{15} and decreases at higher ion doses. In stainless steel again a correlation with the electronegativity was found but just in opposite order compared with Ti[252]. The last example on thermal oxidation is the case of chromium oxidized at 750°C in dry oxygen[253].

Fe, Ne, Mg, S, Ca, Cr, Mn, Fe, Y, Cd, Xe, Cs and Bi have been implanted to depth of up to 400 Å prior to oxidation. In contrast to the systems mentioned above radiation damage effects are the dominating factor for the oxidation behavior of this system. No correlation with the electronegativity and ionic radius could be found but the inert Xe had a distinct inhibiting effect. There is some indication that the oxidation potential is an important quantity but, in general, the oxidation behavior of chromium is not understood.

6.1.2 Aqueous Corrosion

Under aqueous corrosion one understands all phenomena taking place when materials are subjected to aqueous solutions. This includes acids as well as neutral solutions with or without participation of oxygen. The corrosion process in aqueous solutions involves the dissolution of the metal, the hydrogen evolution at the metal and the formation or dissolution of protective oxide layers. The standard method for monitoring these processes is the recording of current-potential relationships (Sect. 4.4.3) either potentiostatically, galvanostatically or potentiokinetically. The latter method yields information on the total range of current density as a function of the electrode potential involving the active-passive transition. Using the former one may study distinct regions with better accuracy. From the measurements one is able to deduce corrosion current densities which are a measure for the corrosion rate, corrosion potentials, critical current densities for passivation and certain indications of the corrosion mechanism.

The greater part of the studies have been performed with iron in neutral solutions. Ashworth et al. examined in detail the behavior of ion-bombarded iron in buffered acetic acid solution (pH = 7.3). At first they looked into the effect caused by argon ion implantation[247]. The potentiokinetic polarization curves showed some differences between the unimplanted and the argon-implanted iron specimen, but only for the first sweep. This indicates the initial presence of an air-formed film which is thicker for the implanted sample than for the untreated iron, as already indicated in the preceding section. Once the film had been removed by cathodic reduction there were no significant differences in the potentiokinetic polarization behavior of the two samples.

This statement is no longer true for iron bombarded with metal ions[254, 255]. The potentiokinetic polarization curves for pure iron and iron bombarded with increasing doses of chromium show a number of differences. In Fig. 31 curves for both materials are plotted in one diagram.

After a first sweep towards the positive which is not shown in the diagram and which is dominated by the dissolution of the airformed oxide layer, a sweep in the positive direction starts at the negative potential end of the cathodic part of the curve. In the first part, from A to the corrosion potential B where the curve becomes anodic, H_2 evolution is the most important process. In this region both samples are very similar. The corrosion potential at B is nearly the same for unimplanted, with Cr^+ implanted and with Ar^+ bombarded iron. From B to C the anodic dissolution of the metal takes place and at C the active to passive transition starts. Here one observes the most significant difference between the two samples. The critical current density for passivation of implanted iron is more than one order of

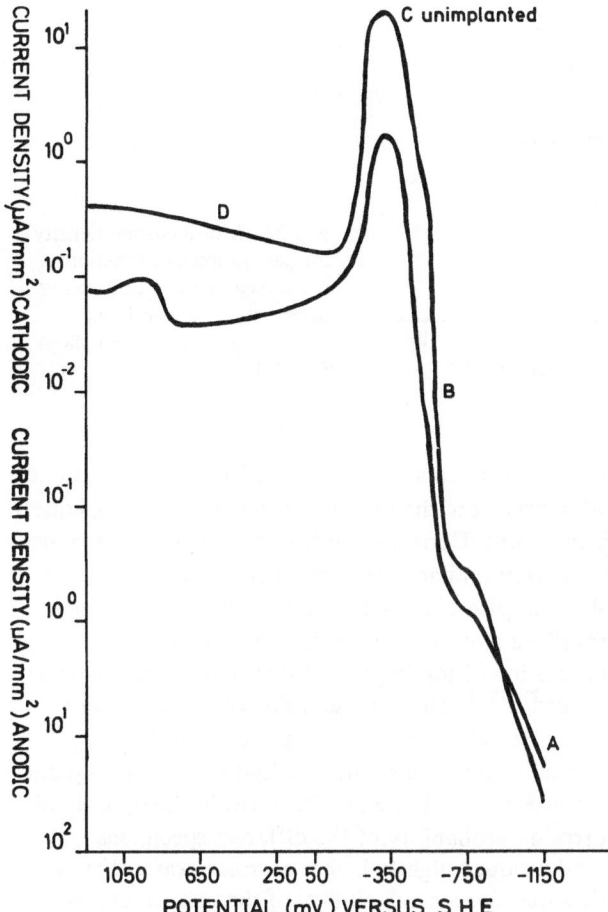

Fig. 31. Potentiokinetic polarization curves for chromium-implanted (5×10^{16} Cr ions/cm^2), in comparison with unimplanted iron (after[254])

magnitude less than for unimplanted iron. This indicates a more pronounced protective effect in the former case. In the last region of the curve (D) the minimum passive current density of the chromium-implanted specimens was still significantly less than that of unimplanted iron.

It is naturally interesting to compare the corrosion behavior of chromium-implanted iron with conventional Fe-Cr alloys.

Fig. 32 shows a comparison of the critical current density for passivation (point C of Fig. 31) plotted against the percentage of Cr in the alloy.

There exists a direct relationship between the two quantities and the calculated compositions of the implanted alloy fit extremely well into the curve. The results indicate that surface alloys produced by ion implantation have electrochemical behavior similar to conventional bulk alloys despite the fact of their thickness being only $10^2 - 10^3$ Å.

Another system under study was Ta$^+$ implanted in iron[256]. Tantalum, because of its neglegible solid solubility in iron, cannot be introduced by conventional alloying techniques. The method used for the analysis of the implanted alloys was, as in

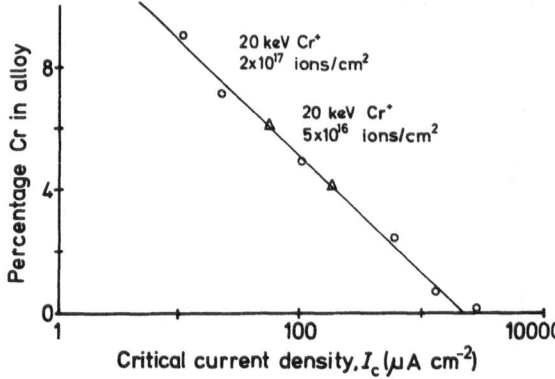

Fig. 32. Critical current density for passivation as a function of percentage of Cr in Cr-Fe alloy (after[255]). o conventional Cr-Fe alloys, △ implanted alloys (Cr⁺ in Fe)

the preceding example, the potentiokinetic polarization. The critical anodic current density for active-passive transition was more than ten times lower for the implanted iron sample than for the unimplanted one. Correspondingly, the corrosion resistance of the tantalum-implanted iron was very similar to that of a conventional Fe-4.9% Cr alloy. Thus, implanted tantalum in spite of its small solid solubility produces a surface alloy layer which behaves like a metastable solid solution of Ta in Fe.

Over the last few years the corrosion of ion-implanted iron in oxygen-free acid solutions has also been investigated[257, 258]. The corrosion rate of the specimens was analyzed by recording the potentiostatic and galvanostatic current density-potential characteristics in the potential region where cathodic hydrogen evolution and anodic dissolution dominate (region A to C in Fig. 31). The corrosion current densities gave an indication of the corrosion probability of the different specimens.

Implantations of rare gases did produce slightly higher corrosion rates than unimplanted iron. It was noteworthy that the reproducibility of the measurements of implanted samples was worse than of unimplanted ones. Oxide layers did not play the important role in this study which they did in the work of Ashworth et al. because in H_2SO_4 solution immediate cathodic reduction of the oxide takes place. Au implantation enhanced the corrosion rate of iron by a factor of ≈ 10; lead implantation inhibited the corrosion rate by the same factor. Copper, finally, did not have much influence, a behavior that is similar to conventional iron-copper alloys with a copper content of less than 10%. These findings proved that ion implantation can vary the corrosion resistance of iron not only in neutral but also in acid solutions.

Further results on the hydrogen evolution reaction in this system are presented in the following chapter.

6.2 Catalysis and Electrocatalysis

Generally speaking, every chemical reaction in which a reaction partner participates without being consumed or formed is a catalytic reaction. Since we are interested in ion-bombardment which is mainly a solid state technique, we can restrict our considerations to two systems:

a) catalytic reactions between gaseous reaction partners at the surface of a solid
   catalyst,
b) electrocatalytic reactions in solutions at the surface of an electrode.

In the following sections we will call the first type of reaction catalysis, the second
type electrocatalysis.

6.2.1 Catalysis and Ion Bombardment

Since about 1930[259 – 261)] a great number of papers concerned with radiation
effects on catalysts have been produced, but very little work in comparison has been
done with ion bombardment. The great majority of the papers dealt with γ-, X-ray,
electron or thermal neutron irradiations. Some of the later studies using protons,
deuterons and a-particles are to some extent comparable to irradiations with ions[262 – 265)]. In some cases enhancement of the catalytic activity was found, in others inhibition or no influence at all. As explanations a number of effects have been dis-
cussed, such as:

geometrical reasons (defect distance)
local or remote electronic defects
point defects
dislocations
surface smoothing
interaction of radiation with surface impurities
surface decomposition.

Only in a few cases was it possible to attribute the chemisorption of a particular
gas to a specific radiation-produced defect. The reader who is interested in radiation
effects on catalysts in general should consult the excellent review of E. H. Taylor[266)].

The attempts to use ion implantation for the modification of catalysts were
mainly restricted to Ar^+ bombardment. Two patents illustrate the effort of finding
industrial applications[267, 268)].

Farnsworth et al. performed a number of experiments in bombarding metals
with low-energy argon ions. The catalytic activity for ethylene hydrogenation was
100 times higher for a freshly bombarded nickel foil than for the same foil after
annealing[269)]. For platinum the enhancement due to the bombardment was only a
factor of 10 and for copper neglegible[270)]. In bombarded germanium no activity
was found[271)]. Ni-Cu alloys showed again an enhancement[272)]. Sosnovsky and co-
workers[273, 274)] bombarded single crystals of silver with argon ions along crystallographic directions exhibiting channeling. In that way the authors claimed to intro-
duce numerous dislocations. All point defects were eliminated by annealing after
irradiation. The catalytic activity of the silver toward formic acid decomposition was
found to be higher after that treatment than in the original crystal.

As far as irradiations with different ions are concerned, the work of Baumgärtner et al.[275, 276)] should be mentioned. They also used ethylene hydrogenation as a
model reaction and were able to show that the bombardment of annealed metal
foils, especially polycrystalline Ni, Co and similar metals, gave rise to a thermal re-
versible activity increase.

In conclusion one may say that it is certainly possible to modify catalysts by using ion implantation. The reasons are mainly physical effects as modifications of the surface or the formation of defects. For the generation of such effects various other methods exist such as cold working, electropolishing or etching. Of much greater interest are chemical effects such as the formation of certain catalytically active alloys which are difficult or impossible to obtain by conventional techniques or the problem of affecting the selectivity of a catalyst in a controlled way.

Whether such chemical effects occur more frequently still seems to be an open question. A major difficulty with studies in this field is the problem of clean surfaces. For obtaining reliable results the work has to be done under ultra-high vacuum conditions. But the findings of such experiments are only of limited value in practice because industrial catalytic processes by no means run under clean conditions.

6.2.2 Electrocatalysis at Ion-implanted Materials

If a catalyst is a substance which alters the rate of a chemical reaction without itself being either consumed or generated in the process, then an electrode which acts as the site for an electrodic reaction and does not dissolve or grow during the process is also a catalyst; it is a charge-transfer catalyst or electrocatalyst[277]. The main difference between a chemical catalyst and an electrocatalyst is the potential dependency of the reaction rate in the latter case. By changing the potential difference across the interface one is able to change the reaction rate by orders of magnitude.

In addition, one needs not to deal with impure surfaces so much because the reaction takes place in solution and surface layers (oxides for example) may dissolve under appropriate potential conditions. Several procedures for achieving activation are also known for electrocatalysts. Ion implantation may be a convenient method of activating and modifying an electrocatalyst. Activation takes place by the introduction of defects and modification by the production of surface-doped layers of varying composition.

Voinov et al. studied the behavior of pyrocarbon in which platinum was implanted[278]. The oxidation of hydrogen and the reduction of oxygen was investigated in $HClO_4$ solution. An increase in H_2 oxidation corresponding to the amount of Pt implanted was observed. The O_2 reduction was not affected by the implantations. In another study difficulties in producing catalytic activity of implanted specimens arose[279]. The bombardment of Cd, Cu and Au with 2×10^{16} Pt ions/cm^2 at 20 keV did not lead to an enhancement of the reaction rate of the hydrogen evolution reaction. On the other hand small amounts of Pt plated on gold produced a considerable effect. Ti implanted with Pt was used as a catalyst for the chlorine evolution reaction in a solution of 2 M NaCl in 10^{-2} M HCl. The bombarded sample produced a current density for the Cl_2 evolution of ≈ 20 mA/cm^2, this being much higher than the value for unimplanted Ti, but more than a factor of 10 lower than for pure Pt[280]. An electrochemical activation step was not necessary for observing the catalytic effect. In contrast, the work of Grenness et al. showed that one sometimes needs electrochemical activation for an ion-implanted catalyst[281] to become effective. The authors implanted monolayer quantities of platinum into tungsten or tungstic oxide and measured the current-voltage characteristics of the implanted electrodes in a solution of H_2SO_4. The current densities for the cathodic hydrogen evolution reaction were several orders of magnitude higher for the implanted anodized tungsten than for pure tungsten and unimplanted anodized tungsten. Implanted pure tungsten also did not show the surprising enhancement of the implanted and anodized tungsten. To reach the high current densities a few initial conditioning voltage cycles were necessary. For the complicated mechanism proposed by the authors the reader should consult the original paper.

Another study dealing with the cathodic hydrogen evolution reaction was concerned with implanted pure iron[257, 258]. The current densities measured for the hydrogen evolution with the implanted iron as cathode in H_2SO_4 solutions are illustrated in Fig. 33 which shows the current-density potential characteristics of pure iron bombarded with $10^{16} - 10^{17}$ Au^{2+}/cm^2 and Pb^{2+}/cm^2. The implantation energy was 400 keV. The iron was irradiated in the form of rotating rods in order to obtain an uniform concentration profile from the surface up to the mean range of the ions.

The figure shows that gold implantation enhances the current density for H_2 evolution by a factor of up to 10; lead implantation, on the other hand, leads to an inhibition. In all cases prior to the measurement the oxide layer was reduced cathodically and the first 100 Å of the specimens dissolved electrochemically. In a second experiment the effect of copper implantation was studied in the same way. In this case no considerable enhancement was measured. Generally speaking, the enhancement or inhibition of the cathodic reaction roughly follows the exchange current densities of the implanted metals. But when the exchange current densities of implant and substrate are close to each other ($\lesssim$ one order of magnitude), additional irradiation effects occur. The exchange current density of copper is about one order

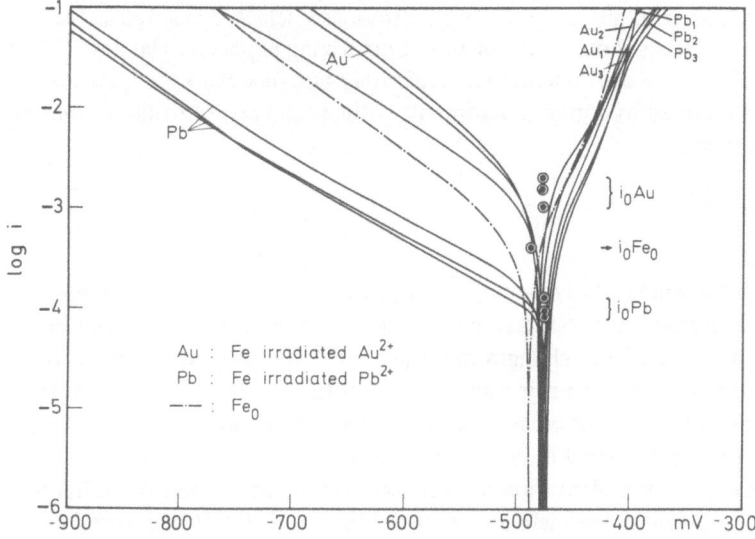

Fig. 33. Cathodic current-potential characteristics (curves on the left) for pure iron, and iron bombarded with Au^{2+} and Pb^{2+} respectively, i_0 are the corresponding corrosion current densities

of magnitude lower than that of iron. Nevertheless, the curves of Cu implanted in iron are generally a little higher than the unimplanted iron. The opposite effect was obtained by Ashworth et al.[256] in their work on aqueous corrosion of iron. Here the cathodic current density for H_2 evolution was 5 times higher for pure iron than for iron implanted with tantalum, despite the exchange current density of tantalum being one order of magnitude higher than that of iron.

One may conclude by saying that it was clearly shown by a number of authors that it is possible to modify to a certain extent the electrocatalytic activity of various substrates by choosing the appropriate elements and conditions for the ion bombardment. In some cases the expected effect did not occur, in others it occurred only after an initial electrochemical activation step.

6.3 Tribological Properties of Ion-implanted Metals

The processing of metal surfaces opens a wide field for applications of ion beam technology. Until now there was not very much activity in this area. Tribology is the science of interacting surfaces in relative motion. Since a number of phenomena like friction, adhesion and wear are interconnected and their interaction forms a tribological system, it is rather difficult to describe it accurately. It seems that the surface asperities and their mechanical properties determine the behavior of two materials under sliding forces. Friction arises from the adhesion at the contact points and sliding occurs under plastic flow. Wear, on the other hand, is defined as the unwanted loss of solid material from sliding surfaces due to their mechanical interaction. Finally, there is no doubt that the hardness of the materials is also of importance.

It is not the purpose of this article to cover the whole field of tribology and ion implantation, but rather to give a short survey with a few selected examples. The reader interested in details should consult the summarizing papers by Hartley[282, 283].

We shall treat the different phenomena separately and deal with the question of whether they are affected by ion bombardment, with special regard to the problems of interest to chemists.

6.3.1 Hardness

The hardness of a surface is relatively easy to measure and it is well known that irradiation with neutrons leads to increased hardness. It is therefore not astonishing that several authors studied the changes in hardness due to ion bombardment. In general nitrogen was used as the projectile and steel as the target[284 – 286]. It was found that in general nitrogen bombardment caused a considerable increase in hardness. The optimal effect occurred at dose values around 5×10^{17} ions/cm^2 and elevated temperature. Boron implantation in iron also resulted in an increase of hardness. Argon as projectile did not yield an effect at $500 - 600°C$[287] in contrast to the situation after implantation at room temperature. The latter effect is plotted in Fig. 34 (left scale) as a function of the dose.

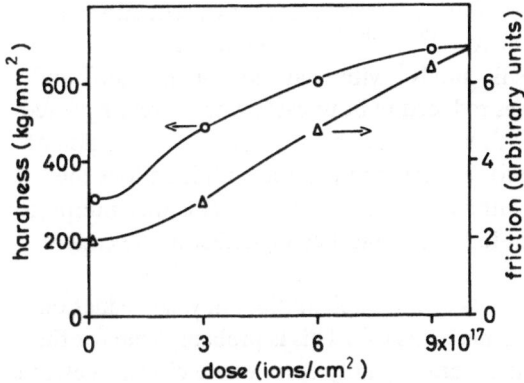

Fig. 34. The variation of the micro-hardness (left curve) and friction co-efficient (right curve) with the dose (40 keV Ar⁺ onto steel) (after[284])

The results are interpreted as showing that the increase in hardness is probably an effect being partly caused by radiation damage, partly by the yield stress coming from the formation of iron nitride or iron boride. It would be very interesting to analyze chemically the quantity of compound formation and to compare this analysis with measurements of hardness.

6.3.2 Friction

The origin of friction is the adhesion between the tips of the asperities of two materials opposite each other, and if one surface is harder than the other, also ploughing of grooves by the harder material.

Fig. 34 shows the relation between hardness (left curve) and friction (right curve) and the effect of ion bombardment on these properties of the material.

Changes in friction may be induced either by variation of the hardness and the surface topography or by changes in the chemical composition of the surface as oxidation or compound formation. The latter provides a kind of lubrication.

The implantation of a considerable dose of metals or non-metals in materials led to large changes of the friction coefficient. The effect of Ar⁺ [284] is certainly mainly due to the creation of a new surface topography or to variations of the natural oxide layer of the material. Non-inert ions as Pb, Mo and Se showed some chemical effect dependent on the type of element introduced. The greatest effect came from lead implantation[288] and was interpreted as a change in the plastic properties of the junction events. Eventually, also the formation of PbO at the surface should be considered.

6.3.3 Wear and Lubrication

The majority of experiments on ion-bombarded materials have been performed using the pin and disc method. Here a standard pin under a load rubs against a rotating disc of the test material. Other methods such as rotating rods or the analysis of the abrasion have been applied less to ion-bombarded samples.

A general result of the experiments, done mainly on steel, was a reduction of the wear after implantation of N^+, C^+ or Ar^+ [284, 288, 283]. A comparison of the effects resulting from light ions such as B^+ and N^+ with heavy ions such as Mo^+ and Ar^+ showed that the wear parameter was reduced in both cases with increasing force on the pin. For materials bombarded with heavy ions this was valid for the region of light and heavy loads on the pin, while for materials implanted with light ions the wear parameter increased again after a minimum at medium load. This was interpreted as a hardening in the case of the light ions and a plasticizing effect in the case of the heavy ions.

Hartley[282, 289] claims "that the effects of ion implantation on wear reduction are due to physical rather than chemical mechanisms". This is probably true for the systems considered in this model. But there are certainly cases where chemical effects are important too, especially in tribological systems, including lubrication. For fluid lubricants it is well known that certain chemical elements improve the adhesion of the lubricating films. These elements may be introduced by ion implantation. This is also possible for high-temperature lubrication where the "lubricant" is incorporated in the material. Here chemical methods for the production of surface films (oxides, fluorides, sulfides) are common. Ion implantation is an alternative method of incorporating the effective element into the surface region of the material.

Summarizing we can make the following statement: Ion implantation is an efficacious method of modifying the tribological behavior of materials. It should be especially interesting when very thin films of the modified material are desirable upon the bulk material, the modified material is not accessible with conventional methods because of a low solubility limit of the system, and a solid lubricant, incorporated in the material with very good adhesion, is necessary.

7 Future Trends

Basic research in ion implantation is slowly moving its emphasis from the semiconductor field to the field of material science, as already mentioned in the introduction. Accordingly, the chemical aspects of ion implantation are gaining more importance and interest. Since the chemical studies represent up to now only a small part of the work done, there is an extraordinarily extensive area of science awaiting future activity. In this short final chapter I shall try to point out the domains of special interest for basic science as well as for applications. The description of future trends is naturally not objective but reflects the personal view of the author.

7.1 Radiation Chemistry

The knowledge about the effects of highly ionizing radiation in *compounds* is very scarce in comparison with the ones of γ-rays or electrons. Systematic investigations of the macroscopic property changes after ion-bombardment with spectroscopic methods would be as interesting as studies on the microscopic defect structure around the paths of individual heavy ions. Mössbauer effect measurements using convenient ions as probes can provide this information. The recording of recovery curves where the defects are created at liquid He temperatures and their behavior measured as a function of increasing temperature seems to be very important. Recovery curves are well known for metals and help to understand the phenomena connected to the transition from disorder to order. Besides this, the radiation chemistry of ion-bombarded compounds is relevant for the questions of radiation effects in nuclear power plants and fusion devices, the interactions of the solar wind and the cosmic rays in space, on the moon and in the atmosphere.

7.2 Solid State Chemistry

Up to now knowledge of the chemical state of energetic ions or atoms in solids results entirely from measurements at room temperature. From these measurements one gets evidence that a considerable thermal recovery process takes place drastically changing the situation immediately after implantation. Therefore, experiments at liquid He temperatures would be of great interest. Such experiments have some similarity with matrix isolation techniques: in the case of matrix isolation interactions of initially single atoms or molecules of two species in an inert matrix at low temperatures as a function of temperature are investigated. In the ion implantation experiment one studies, as a function of temperature, the interaction of single implanted atoms and defects or fragments produced during the implantation process in a matrix which is considered as an ideal crystal and does not take part in the reaction. To give an example: an alkaline earth ion implanted at low temperatures in NaCl will react mainly with the vacancies, electrons, Cl-Ions or atoms in interstitial positions and Cl_2-molecules or molecular ions created during the slowing down process. These different interactions are strongly temperature dependent.

Another interesting field will be the formation of unconventional compounds arising from the interactions mentioned above. Since these compounds are mostly unstable at RT, their production requires working under low-temperature conditions. The analysis must also be performed at low temperatures, thus spectroscopic measurements done on-line with ion implantation are needed.

Finally, the formation of compounds during the bombardment of metals and thin films with reactive non-metals is an unexplored field with immediate consequences for practical applications. The formation of these compounds requires usually a certain minimum dose and decreases again above a maximum dose. Unfortunately, the experimental material illustrating these relations is extremely rare. Precise experiments using spectroscopic methods are necessary to identify the compounds unambiguously and to replace speculations by experimental facts. In this connection the interaction between energetic hydrogen and the materials in discussion for future use in fusion devices or the ones on the surface of the moon and in space should be given particular reference.

7.3 Corrosion and Oxidation

The investigations of corrosion of ion-bombarded metals have arrived at a point where one has a rough idea about what is going on. For a number of systems it is possible to predict which element can be implanted to improve corrosion resistance. The future point in question is to what extent the implantation method will find practical application. There is no doubt that it will never be a routine method for cheap or voluminous materials or workpieces. On the other hand, it may be a suitable method for small expensive parts or in cases where the replacement of parts is difficult or costly.

The implantation method will always be superior when very thin protective surface layers are needed or if one wants to form "alloys" for which low solubility limits forbid the use of conventional techniques. The work done until now has shown that implantation is a very good method to do basic research in the field of corrosion and oxidation. The fact of allowing every possible combination of surface alloys and a very low and controlled concentration of impurities as well as studying the influence of defects is unique. Therefore, this method should play a more and more important role in corrosion science in the future.

7.4 Catalysis and Electrocatalysis

The situation regarding heterogeneous catalysis is rather confused. More fundamental research is needed to clarify the problem of whether certain elements, when implanted, cause specific chemical effects other than radiation damage. Moreover, it is an open question as to whether it is possible to vary the selectivity of a catalyst. Another difficulty is the standardization of the surfaces under study. Either one works under UHV conditions which have little relationship to practical applications, or under "normal" conditions which means that surface contaminations are unavoidable.

In the field of electrocatalysis the situation seems to be somewhat better as far as the problems of clean surfaces and the existence of chemical effects are concerned. Unfortunately, so far only a few reactions have been studied. Thus, much more systematic fundamental research has to be done. From the results already available one can extract the hope that ion-bombardment will be of future importance for fields involving electrocatalytical reactions such as, for example, hydrogen technology, energy conversion, fuel cells and electrochemical redox reactions.

7.5 Mechanical Properties of Surfaces

For ion-beam modifications of the mechanical surface properties the same considerations are valid as for the modification of corrosion behavior. Their applications in the treatment of materials will be restricted to special cases where the costs are not of main importance.

It is well established now that ion-bombardment can improve the hardness of materials or the resistance against wear. The question of improving the adhesion of lubricants or of producing self-lubricating thin films on the surface must be subject to further investigations.

Another field for future activity is the influence of ion implantation on the properties of surface coatings. It should be possible to affect the conditions of the electrolytical coating process as well as the mechanical characteristic of coatings by modifying the substrate by means of ion beams.

8 References

1. Picraux, S. T., EerNisse, E. P., Vooke, F. L. (eds.): Applications of ion beams to metals, New York: Plenum Press, 1974
2. Picraux, S. T.: Applications of ion beams to materials, Carter, G., Colligon, J. S., Grant, W. A. (eds.). London: Inst. Physics 1976, p. 183
3. Franklin, J. L. (ed.): Ion molecule reactions. London: Butterworth 1972
4. McDaniel, E. W. et al.: Ion molecule reactions. New York: Wiley-Interscience 1970
5. Bohr, N.: Phil. Mag. 25, 10 (1913)
6. Lindhard, J., Scharff, M., Schiott, H. E.: Kgl. Danske Vid. Selsk. Mat., Fys. Medd. 33, 14 (1963)
7. Bohr, N., Lindhard, J.: Kgl. Danske Vid. Selsk. Mat., Fys. Medd. 28, 7 (1954)
8. Bethe, H. A.: Ann. Phys. 5, 325 (1930)
9. Inokuti, M.: Rev. Mod. Phys. 43, 297 (1971)
10. Bloch, F.: Ann. Physik 16, 285 (1933)
11. Fano, U.: Ann. Rev. Nucl. Sci. 13, 1 (1963)
12. Bloom, D., Sautter, G. D.: Phys. Rev. Lett. 26, 607 (1971)
13. Lindhard, J., Scharff, M.: Phys. Rev. 124, 128 (1961)
14. Bohr, N.: Kgl. Dan. Vid. Selsk. Mat., Fys. Medd. 18, 8 (1948)
15. Townsend, P. D., Kelly, J. C., Hartley, N. E. W.: Ion implantation, sputtering and their applications. London: Academic Press 1976, p. 14
16. Winterbon, K. B., Sigmund, P., Sanders, J. B.: Kgl. Dan. Vid. Selsk. Mat., Fys. Medd. 37, 14 (1970)
17. Lehmann, Ch.: Interaction of radiation with solids. Amsterdam: North-Holland 1977, p. 107
18. Schiott, H. E.: Rad. Eff. 6, 107 (1970)
19. Dearnaley, G., Freeman, J. H., Nelson, R. S., Stephan, J.: Ion implantation. Amsterdam: North-Holland 1973, p. 766 ff.
20. Gibbons, J. F., Johnson, W. S., Mylroie, S. W.: Projected Range Statistics. New York: Wiley-Interscience (1975)
21. Wilson, R. G., Brewer, G. R.: Ion beams. New York: Joh. Wiley 1973, p. 353 ff.
22. Townsend, P. D., Kelly, J. C., Hartley, N. E. W.: Ion implantation, sputtering and their applications. London: Academic Press 1976, p. 31
23. Jespersgaard, P., Davies, J. A.: Can. J. Phys. 45, 2983 (1967)
24. Townsend, P. D., Kelly, J. C., Hartley, N. E. W.: Ion implantation, sputtering and their applications. London: Academic Press 1976, p. 37
25. Kornelsen, E. V., Brown, F., Davies, J. A., Domeij, B., Piercy, G. R.: Phys. Rev. A. 136, 849 (1964)
26. Williams, J. S.: Phys. Letters 51 A, 85 (1975)
27. Brice, D. K.: Rad. Eff. 6, 77 (1970)
27a. Oen, O. S., Robinson, M. T.: J. Appl. Phys. 35, 2515 (1964)

28. Townsend, P. D., Kelly, J. C., Hartley, N. E. W.: Ion implantation, sputtering and their applications. London: Academic Press 1976, p. 41
29. Piercy, G. R., Cargo, M. Mc., Davies, J. A.: Can. J. Phys. *42*, 1116 (1964)
30. Nelson, R. S., Thompson, M. W.: Phil. Mag. *8*, 1677 (1963)
31. Almen, O., Bruce, G.: Nucl. Inst. and Meth. *11*, 257 (1961)
32. Sigmund, P.: Phys. Rev. *184*, 383 (1969)
33. McDonald, R. J.: Adv. Physics *19*, 457 (1970)
34. Thompson, M. W.: Defects and radiation damage in metals. Cambridge: University Press 1969
35. Rosenbaum, H. S.: Treatise on material science and technology. New York: Academic Press 1975, p. 19
36. Brinkman, J. A. in: Physics of solids. Billington, D. S. (ed.) New York: Academic Press 1963, p. 830
37. Thompson, M. W.: Defects and radiation damage in metals. Cambridge: University Press 1969, p. 303
38. Hemmerich, J.: IAEA-171, Vienna, p. 77 ff. (1975)
39. Schilling, W., Sonnenberg, K.: J. Phys. F, *3*, 322 (1973)
40. Sonder, E.: J. Physique *34*, C9, 483 (1973)
41. Hayes, W.: J. Physique *34*, C9, 499 (1973)
42. Davies, J. A., Jespersgaard, P.: Can. J. Phys. *44*, 1931 (1966)
43. Rühle, M. R.: Radiation Damage in Reactor Materials, Vol. I, IAEA, Vienna, p. 113 (1969)
44. Hull, D.: Introduction to dislocations. Oxford: Pergamon Press 1975, p. 20
45. Nelson, R. S., Mazey, D. J.: Radiat. Effects *18*, 127 (1973)
46. Kaminsky, M., Das, S. K.: Appl. Phys. Lett. *23*, 293 (1973)
47. Dearnaley, G., Freeman, J. H., Nelson, R. S., Stephen, J.: Ion implantation. Amsterdam: North Holland 1973
48. Wagner, H., Walcher, W. (eds.): Proc. Intern. Conf. Electromagnetic Isotope Separators, Marburg, 1970
49. Proc. Intern. Conf. Low Energy Ion Beams, Salford 1977
50. Viehböck, F., Winter, H., Bruck, M. (eds.): Proc. 2nd Intern. Conf. Ion Sources, Vienna 1972
51. Langmuir, F., Taylor, J. B.: Phys. Rev. *34*, 423 (1933)
52. Nielsen, K. O.: Nucl. Instr. and Methods *1*, 289 (1957)
53. Kirchner, R.: Intern. Conf. Low Energy Ion Beams, Salford, U.K. September 5 – 8, 1977
54. Johnson, P. G., Bolson, A., Henderson, C. M.: Nucl. Instr. and Methods *106*, 83 (1973)
55. Wolf, B. H.: Nucl. Instr. and Methods *139*, 13 (1976)
56. Sidenius, G., Skilbreid, O., in: Electromagnetic separation of radioactive isotopes. Higatsberger and Viehböck (eds.) Wien: Springer 1961
57. Freeman, J. H., Sidenius, G.: Nucl. Instr. Methods *107*, 477 (1973)
58. Freeman, J. H., in: Ion implantation, Amsterdam: North Holland 1973, p. 454 ff.
59. Morgan, D. V. (ed.): Channeling. London: John Wiley 1973
60. Wolf, G. K.: Nucl. Instr. and Methods *139*, 147 (1976)
61. Freeman, J. H.: cited in [47], p. 423 (1970)
62. Freeman, J. H.: AERE Report R 6138 (1969)
63. Dearnaley, G., Freeman, J. H., Nelson, R. S., Stephen, J.: Ion implantation. Amsterdam: North Holland 1973, p. 416f
64. Nelson, R. S., in: Physics of ionized gases. Navincek, B. (ed.). Ljubljana 1976, p. 421
65. Yurasova, V. E., in: Physics of ionized gases. Navincek, B. (ed.). Ljubljana 1976, p. 493
66. Mayer, J. W., Eriksson, L., Picraux, S. T., Davies, J. A.: Can. J. Phys. *46*, 663 (1968)
67. Grant, W. A., Williams, J. S., Christodoulides, C. E., in: Physics of ionized gases. Navincek, B. (ed.). Ljubljana 1976, p. 340
68. Dearnaley, G., Freeman, J. H., Nelson, R. S., Stephen, J.: Ion implantation. Amsterdam: North Holland 1973, p. 9 ff.
69. Nelson, R. S., in: Channeling. Morgan, D. V. (ed.). New York: Wiley-Interscience 1973, p. 259

70. Lee, Y. A., Brosious, P. R., Cheng, L. J., Corbett, J. W., in: Ion implantation in semiconductors. Namba, S. (ed.). New York: Plenum Press 1975, p.519
71. Shimizu, T., Hasegawa, S., Karimoto, H., in: Ion implantation in semiconductors. Namba, S. (ed.). New York: Plenum Press 1975, p.525
72. Murakami, K., Masuda, K., Gamo, K., Namba, S., in: Ion implantation in semiconductors. Namba, S. (ed.). New York: Plenum Press 1975, p.533
73. Henderson, B.: Defects in crystalline solids. London: Edward Arnold 1972
74. McGill, J. C., Kurtin, S. L., Shifrin, G. A.: J. Appl. Phys. *41*, 246 (1970)
75. Crowder, B. L., Smith, J. E., Brodsky, M. H., Nathan, M. I.: Ion implantation in semiconductors. Ruge, J., Graul, J. (eds.). New York, Heidelberg, Berlin: Springer 1971, p.255
76. Gruen, D. M., Varma, R., Wright, R. B.: J. Chem. Phys. *64*, 5000 (1976)
77. Wright, R. B., Gruen, D. M.: Rad. effects *33*, 133 (1977)
78. DeWaard, H., in: Mössbauer spectroscopy and its applications. IAEA, Vienna 1972, p.123
79. Thompson, M. W.: Defects and radiation damage in metals. Cambridge: University Press 1969, p.257 ff.
80. Ion Beam Handbook. J. W. Mayer, R. Rimini. New York: Acad. Press, p. 80 (1977)
81. Davies, J. A., in: Channeling. Morgan, D. V. (ed.). London: John Wiley 1973, p.392 ff.
82. DeWaard, H., Feldman, L. C., in: Applications of ion beams to metals. Picraux, S. T., Eer Nisse, E. P., Vook, F. L. (eds.). New York: Plenum Press 1974, p.37
83. Mayer, J. W., Turos, A.: Thin Solid Films *19*, 1 (1973)
84. Chenin, J. F., Mitchell, J. V., Saris, F. W., in: Ion implantation in semiconductors. Ruge, J., Graul, J. (eds.). New York, Heidelberg, Berlin: Springer 1971, p.295
85. Borders, J. A., Beezhold, W., in: Ion implantation in semiconductors. Ruge, J., Graul, J. (eds.). New York, Heidelberg, Berlin: Springer 1971, p. 241
86. Ranon, U., Yaniv, A.: Physics Letters *9*, 17 (1974)
86a Newman, R. C., Woodward, R. J.: J. Phys. C, *7*, L432 (1974)
87. Gütlich, Ph.: Chemie in unserer Zeit *4*, 133 (1970), *5*, 131 (1971)
88. Goldanskii, V. I., Herber, R. H. (eds.): Chem. appl. of Mössbauer spectroscopy. New York: Academic Press 1968
89. Gütlich, P., Link, R., Fritsch, T., Wolf, G. K.: Nucl. Instr. Meth. *148*, 573 (1978)
90. Drentje, S. A., Reintsema, S. R.: Nucl. Instr. Meth. *133*, 421 (1976)
91. Maddock, A. G.: Int. Rev. Sci., Inorg. Chem. Ser. *1, 8*, 213 (1972)
92. Stanek, J., Sawicki, J., Sawicka, B.: Nucl. Instr. Meth. *130*, 613 (1975)
93. Sawicki, J. A., Sawicka, B. D., Stanek, J.: Nucl. Instr. Meth. *138*, 565 (1976)
94. Stöcklin, G.: Chemie heißer Atome. Weinheim: Verlag Chemie 1969
95. Müller, H.: Angew. Chem. *79*, 128 (1967)
96. Maddock, A. G.: Intern. Rev. Sci., Inorg. Chem. Ser., Radiochemistry *8*, 273 (1975)
97. Szilard, L., Chalmers, T. A.: Nature *134*, 462 (1934)
98. Baumgärtner, F., Reichold, P., in: Chemical effects of nuclear transformations II 319. Vienna 1961, p.196
99. Wolf, G. K., Fröschen, W., Sahm, U., in: Ion implantation in semiconductors. Namba, S. (ed.). New York: Plenum Press 1975, p.317
100. Townsend, P. D., Kelly, J. C., Hartley, N. E. W.: Ion implantation, sputtering and their applications. London: Academic Press 1976, p.97
101. Bogh, E., in: Channeling. Morgan, D. V. (ed.). London: Wiley 1973, p.435
102. Bradford, P. M., Case, B., Dearnaley, G., Turner, J. F., Woolsey, I. S.: Corr. Sci. *16*, 747 (1976)
103. Cairns J. A., Marwick, A. D., Mitchell, I. V.: Thin Solid Film *19*, 91 (1973)
104. Cairns, J. A., Surface Sci. *34*, 638 (1973)
105. Siegbahn, K., Nordling, C., Fahlmann, A., Nordberg, R., Hamrin, K., Hedman, J., Johansson, G., Bergmark, T., Karlsson, S., Lindgren, J., Lindberg, B.: ESCA, Almquist and Wiksells, Uppsala 1967
106. Staib, P.: Rad. Effects *18*, 217 (1973)
107. Ertl, G., Küppers, J.: Low energy electrons and surface chemistry. Weinheim: Verlag Chemie 1974

108. Evans, C. A., Jr.: Thin Solid Films *19*, 11—19 (1973)
109. Benninghoven, A.: Z. Naturforsch. *22a*, 841 (1967)
110. Barber, M., Vickermann, J. C., in: Surface and defect properties of solids, 5. London: Chem. Soc. 1976, p. 162
111. Whitton, J. L., in: Channeling. Morgan, D. V. (ed.). London: John Wiley 1973, p. 225
112. Townsend, P. D., Kelly, J. C., Hartley, N. E. W.: Ion implantation, sputtering and their applications. London: Academic Press 1976, p. 188
113. Becker, M., Heilgeist, M., Wolf, G. K., in: Low energy ion beams. Stephens, K. G., Wilson, I. H., Moruzzi, J. L. (eds.). London: Inst. Physics 1977, p. 185
114. Ashworth, V., Grant, W. A., Procter, R. P. M.: Corr. Sci. *16*, 661 (1976)
115. Ferber, H.: private communication (1977)
116. Vetter, K. J.: Angew. Chem. *13*, 277 (1961)
117. Lorenz, W. J., Eichkorn, G., Mayer, C.: Corr. Sci. *1*, 357 (1967)
118. Bockris, J. O. M., Reddy, A. K. N.: Modern electrochemistry. New York: Plenum Press 1970
119. Johnson, E. R.: Irradiation-induced decomposition of inorganic molecular ions. New York: Gordon and Breach 1970
120. Drigo, A. V., LoRusso, S., Mazzoldi, P., Goode, P. D., Hartley, N. E. W.: Rad. Effects *33*, 161 (1977)
121. Naguib, H. M., Singleton, J. F., Grant, W. A., Carter, G.: J. Mat. Science *8*, 1633 (1973)
122. Morhange, J. F., Beserman, R., Bourgoin, J. C., Brosious, P. R., Lee, Y. H., Cheng, L. J., Corbett, J. W., in: Ion implantation in semiconductors. Namba, S. (ed.). New York: Plenum Press 1975, p. 457
123. Olley, J. A., Yoffe, A. D., in: Ion implantation in semiconductors. Ruge, J., Graul, J. (eds.). New York, Heidelberg, Berlin: Springer 1971, p. 248
124. Hutchby, J. H., Webster, R. C., Miller, W. E., in: Ion implantation in semiconductors. Ruge, J., Graul, J. (eds.). New York, Heidelberg, Berlin: Springer 1971, p. 385
125. Miller, W. E., Hutchby, J. A., Webster, R. C., in: Ion implantation in semiconductors. Ruge, J., Graul, J. (eds.). New York, Heidelberg, Berlin: Springer 1971, p. 373
126. Hutchby, J. A.: Rad. effects *16*, 189 (1972)
127. Walsh, W.: Solid State Electr. *20*, 813 (1977)
128. Gruen, D. M., Varma, R., Wright, R. B.: Chem. Phys. *64*, 5000 (1976)
129. Mohs, E.: Dissertation, Univers. Heidelberg 1976
130. Sahm, U.: Dissertation, Univers. Heidelberg 1976
131. Wolf, G. K., Fröschen, W., Sahm, U., in: Ion implantation in semiconductors. Namba, S. (ed.). New York: Plenum Press 1975, p. 317
132. Fleisch, J., Gütlich, P., Mohs, E., Wolf, G. K.: 7th Intern. Hot Atom Chemistry Symposium, Jülich, Sept. 1973 AED-Conf., 13402 — 080
133. Link, R., Gütlich, P., Fritsch, T., Wolf, G. K.: unpublished
134. Sonder, E.: J. Phys. *C-9, 34*, 483 (1973)
135. Whitton, J. L., in: Channeling. Morgan, D. V. (ed.). London: John Wiley 1973, p. 248
136. Furukawa, S., Ishiwara, H.: Japan. J. Appl. Phys. *11*, 1062 (1972)
137. Wehner, G. K., Kenknight, C. E., Rosenberg, D.: Planet. Space Sci. *11*, 1257 (1963)
138. Oechsner, H., in: Physics of ionized gases. Navincek, B. (ed.). Ljubljana: J. Stefan Inst. 1976, p. 477
139. Naguib, H. M., Kelly, R.: Rad. Effects *25*, 1 (1975)
140. Biersack, J. P., Santner, E.: Nucl. Instr. Meth. *132*, 229 (1976)
141. Kelly, R., Sam, N. Q.: Rad. Effects *19*, 39 (1973)
142. Murti, D. K., Kelly, R.: Surface Sci. *47*, 282 (1975)
143. Parker, Th., Kelly, R.: J. Phys. Chem. Sol. *36*, 377 (1975)
144. Tarney, M. L., Wehner, G. K.: J. Appl. Phys. *43*, 2268 (1972)
145. Oechsner, H., Gerhard, W.: Surface Sci. *44*, 480 (1974)
146. Harbottle, G.: Ann. Rev. Nucl. Sci. *15*, 89 (1965)
147. Wolfgang, R.: Ann. Rev. Nucl. Sci. *16*, 15 (1965)
148. Wiles, D. R., Baumgärtner, F.: Fortschritte der chem. Forschung *32*, 64 (1972)

149. Andersen, T., Sørensen, G., in: Electromagnetic isotope separators and their applications. Koch, J., Nielsen, K. O. (eds.). Amsterdam: North-Holland 1965
150. Andersen, T., Sørensen, G.: Trans. Faraday Soc. *62*, 3427 (1966)
151. Cogneau, M., Duplatre, G., Vargas, J. I.: J. Inorg. Nucl. Chem. *34*, 3021 (1972)
152. Freeman, J. H., Kasrai, M., Maddock, A. G.: Chem. Commun. 979 (1967)
153. Andersen, T., Ebbesen, A.: Trans. Faraday Soc. *67*, 3540 (1971)
154. Kasrai, M., Maddock, A. G., Freeman, J. H.: Trans. Faraday Soc. *67*, 2108 (1971)
155. Maddock, A. G.: Intern. Rev. Sci. Inorg. Chem. *8*, 317 (1975)
156. Mohs, E., Wolf, G. K., Wagner, U., Wagner, F. E.: Int. Conf. Mössbauer spectroscopy, Bucharest, Romania 5–10 Sept. 1976
157. Wolf, G. K.: Nucl. Instr. Meth. *139*, 147 (1976)
158. Fleisch, J., Gütlich, P., Mohs, E., Wolf, G. K.: Int. Conf. Mössbauer spectroscopy. Cracow, 26–30 July, 1975
159. Gruen, D. M., Siskind, B., Wright, R. B.: J. Chem. Phys. *65*, 363 (1976)
160. Roy, R., Greer, R. T.: Solid State Commun. *5*, 109 (1967)
161. Drigo, A. V., LoRusso, S., Mazzoldi, P., Goode, P. D., Hartley, N. E. W.: Rad. Effects *33*, 161 (1977)
162. Arnold, G. W., Borders, J. A., in: Application of ion beams to materials. Carter, G., Colligon, J. S., Grant, W. A. (eds.). London: Inst. Physics 1976, p. 128
163. Andersen, T., Langvad, T., Sørensen, G.: Nature *218*, 1158 (1968)
164. Yoshihara, K., Kishimoto, M., Takahashi, M., Suzuki, S., Shiokawa, T.: Radiochimica Acta *21*, 148 (1974)
165. Sahm, U.: Thesis, Universität Heidelberg 1976
166. Wolf, G. K., Sahm, U., Mohs, E., in: Physics of ionized gases. Navincek, B. (ed.). Ljubljana: J. Stefan Institute 1976, p. 266
167. Jenkins, G. M., Wiles, D. R.: Ann. Conf., Chem. Inst. Canada, Quebec 7.6.1972
168. Jenkins, G. M., Wiles, D. R.: Chem. Commun. 1177 (1972)
169. Kanellakopulos-Drossopulos, W., Wiles, D. R.: J. Inorg. Nucl. Chem. *38*, 947 (1976)
170. Gromova, J. J., Islamova, K. H. M., Lebedev, N. A., Zaitseva, N. G.: J. Inorg. Nucl. Chem. *38*, 961 (1976)
171. Akcay, H., Cailleret, J., Paulus, J. M., Abbe, J. Ch.: 7th Intern. Hot Atom Chem. Symposium Jülich, 10–14. Sept. 1973
172. Lemmon, R. M.: 7th Intern. Hot Atom Chem. Symposium Jülich, 10–14 Sept. 1973
173. Wolf, G. K., Fritsch, T.: Radiochimica Acta *11*, 195 (1969)
174. Rauscher, H., Sutin, N., Miller, J. M.: J. Inorg. Nucl. Chem. *12*, 378 (1960)
175. Wolf, G. K.: Radiochimica Acta *6*, 39 (1966)
176. Mohs, E.: Thesis, Univers. Heidelberg 1976
177. Rössler, K.: Thesis, Univers. Köln 1967
178. Wolf, G. K.: Chemiker-Z. *99*, 362 (1975)
179. Fleisch, J., Gütlich, P., Mohs, E., Wolf, G. K.: Int. Conf. on Mössbauer Spectroscopy, Cracow, 26–30 July 1975
180. Fleisch, J.: Thesis, Darmstadt 1976
181. Fenger, J., Maddock, A. G.: J. Chem. Soc. (A) 3255 (1970)
182. Siekierska, K. E., Fenger, J., Olsen, J.: J. Chem. Soc. (D), 2020 (1972)
183. Rössler, K., Pross, L.: Radiochem. Radioanalyt. Letters *18*, 291 (1974)
184. Rössler, K., Pross, L.: 9th Intern. Hot Atom Chemistry Symposium, Blacksburg, Virginia 1977
185. Ion implantation in semiconductors. Ruge, J., Graul, J. (eds.). New York, Heidelberg, Berlin: Springer 1971
186. Ion implantation in semiconductors. Namba, S. (ed.). New York: Plenum Press
187. Nelson, R. S., in: Physics of ionized gases. Navincek, B. (ed.). Ljubljana: J. Stefan Institute. 1976, p. 421
188. DeWaard, H., Drentje, S. A.: Physics. Lett. *20*, 38 (1966)
189. DeWaard, H., Heberle, J., Schurer, P. J., Hasper, H., Koks, F. W. J.: Hyperfine structure and nuclear radiations. Amsterdam: North Holland 1968, p. 331

190. DeWaard, H., Kemerink, G.: Int. Conf. Applications of the Mössbauer Effect, Corfu, 13–17 Sept. 1976
191. Cohen, R. L., Beyer, G., Deutch, B.: Phys. Rev. Lett. *33*, 518 (1974)
192. Alexander, R. B., Ansaldo, E. J., Deutch, B. I., Gellert, J., in: Applications of ion beams to metals. Picraux, S. T., EerNisse, E. P., Vook, F. L. (eds.). New York: Plenum Press 1974, p. 365
193. Niesen, L., Kikkert, P. J.: Intern. Conf. Hyperfine Interactions Studied in Nucl. Reactions and Decay, Uppsala, Sweden, p. 160, (1974)
194. Thomé, L., Bernas, H., Chaumont, J., Abel, F., Bruneaux, M., Cohen, C.: Physics Lett. *54A*, 37 (1975)
195. Kurup, M. B., Prasad, K. G., Sharma, K. P.: Proc. of the Nucl. Phys., Sol. State Phys. Symp., Bombay p. 284 (1976)
196. McCracken, G. M., Erents, S. K., in: Applications of ion beams to metals. Picraux, S. T., EerNisse, E. P., Vook, F. L. (eds.). New York: Plenum Press 1974, p. 585
197. Das, S. K., Kaminsky, M., in: Applications of ion beams to metals. Picraux, S. T., EerNisse, E. P., Vook, F. L. (eds.). New York: Plenum Press 1974, p. 543
198. Bohdansky, J., Roth, J., Poschenrieder, W. P., in: Applications of ion beams to materials. Carter, G., Colligon, J. S., Grant, W. A. (eds.). London: Inst. Physics 1976, p. 307
199. Erents, S. K., in: Applications of ion beams to materials. Carter, G., Colligon, J. S. Grant, W. A. (eds.). London: Inst. Physics 1976, p. 318
200. McCracken, G. M.: Rep. Progr. Phys. *38*, 311 (1975)
201. Gruen, D. M., in: Chemistry of fusion technology. Gruen, D. M. (ed.). New York: Plenum Press 1972, p. 229
202. Sheft, I., Reis, A. H., Jr., Gruen, D. M., Peterson, S.: Trans. Am. Nucl. Soc. *22*, 166 (1975)
203. Pavlov, P. V., Zorin, E. I., Tetelbaum, D. I., Lesnikov, V. P., Ryzhkov, G. M., Pavlov, A. V.: Phys. Stat. Solidi (a) *19*, 373 (1973)
204. Deery, M., Goh, K. H., Stephens, K. G., Wilson, I. H.: Thin Solid Films *17*, 59 (1973)
205. Goh, K. H., Stephens, K. G., Wilson, I. H., in: Ion implantation in semiconductors. Namba, S. (ed.). New York: Plenum Press 1975, p. 325
206. Wilson, I. H., Goh, K. H., Stephens, K. G.: Thin Solid Films *33*, 205 (1976)
207. Watanabe, M., Tohi, A.: Japan. J. Appl. Phys. *5*, 737 (1966)
208. Pavlov, P., Shitova, E.: Sov. Phys. Dokl. *12*, 11 (1967)
209. Bularin, J. M., Otto, G., Storbeck, I., Schenk, M., Wagner, H.: Thin Solid Films *4*, 255 (1959)
210. Lesheiko, L. V., Lubopytovo, E. V.: Proc. Intern. Conf. Ion Beam Modific. Mat., Budapest 1978
211. Edelmann, F. L., Kuznetsov, O. N., Lezheiko, L. V., Lubopytova, E. V.: Rad. Effects *29*, 13 (1975)
212. Gruen, D. M., Varma, R., Wright, R. B.: Chem. Phys. *64*, 5000 (1976)
213. Gruen, D. M., Siskind, B., Wright, R. B.: J. Chem. Phys. *65*, 363 (1976)
214. Chang Tung-ho, Teng Hsien-tsan, in: Applications of ion beams to materials. Carter, G., Colligon, J. S., Grant, W. A. (eds.). London: Inst. Physics 1975, p. 96
215. Siskind, B., Gruen, D. M., Varma, R.: J. Vac. Sci. Technol. *14*, 537 (1977)
216. Picraux, S. T., Vook, F. L.: Phys. Rev. B, to be published
217. Rosner, D. E., Allendorf, H.: Proc. Int. Conf. Heterogenic Kinetics at Elevated Temp. University of Pennsylvania, p. 231 (1969)
218. Coulon, M., Bonnetain, L.: J. Chim. Phys. *71*, 711, 725 (1974)
219. McCracken, G. M., Partridge, J. W.: Proc. Int. Conf. Surface Effects in Fusion Devices San Francisco 1976
220. Blauth, E. W., Meyer, E. H.: Z. Angew. Physik 19, 549 (1965)
221. Erents, S. K., Braganza, C. M., McCracken, G. M.: Proc. Int. Conf. Surface Effects in Fusion Devices. San Francisco 1976
222. Roth, J. J., Bohdansky, J., Poschenrieder, W., Sinka, M. K.: Proc. of the Int. Conf. Surface Effects in Fusion Devices. San Francisco 1976

223. McCracken, G. M.: Physics of ionized gases. Navincek, B. (ed.). Ljubljana: J. Stefan Inst. 1976, p. 409
224. Balooch, M., Olander, D. R.: J. Chem. Phys. *63*, 4772 (1975)
225. Hinterberger, H., Weber, H. W., Voshage, H., Wänke, H., Begemann, F., Wlotzka, F.: Proc. Apollo 11 Lunar Sci. Conf., Geochim. Cosmochim. Acta, Suppl. 1, *2*, 1269 (1970)
226. Müller, O.: Habilitationsschrift, Univers. Heidelberg 1976
227. Müller, O.: Proc. Third Lunar Sci. Conf. Geochim. Cosmochim. Acta, Suppl. 3, *2*, 2059 (1972)
228. Pillinger, C. T., Cadogan, P. H., Maxwell, J. R., Eglinton, G., Mays, B. J., Grant, W. A., Nobes, M. J.: Nature *235*, 108 (1972)
229. Bibring, J. P., Burlingame, A. L., Chaumont, J., Langeven, Y., Maurette, M., Wszolek, P. C.: Proc. Fifth Lunar Sci. Conf., Geochim. Cosmochim. Acta, Suppl. 5, *2*, 1747 (1974)
230. Chang, S., Mack, R., Gibson Jr., E. K., Moore, G. W.: Proc. Fourth Lunar Sci. Conf., Geochim. Cosmochim. Acta, Suppl. 4, *2*, 1509 (1973)
231. Chang, S., Lennon, K.: Proc. Sixth Lunar Sci. Conf., Geochim. Cosmochim. Acta, Suppl. 6, *2*, 2179 (1975)
232. Baur, H., Frick, U., Funk, H., Schultz, L., Sigrer, P.: Proc. Third Lunar Sci. Conf., Geochim. Cosmochim. Acta, Suppl. 3, *2*, 1947 (1972)
233. Ducati, H., Kalbitzer, S., Kiko, J., Kirsten, T., Müller, H. W.: The Moon *8*, 210 (1973)
234. Conf. Ion Implantation and Ion Beam Analysis in Corrosion, Manchester, 8–10 April 1975
235. Conf. Ion Implantation and Ion Beam Analysis Techniques in Corrosion, Manchester, 28–30 June 1978
236. Goode, P. D., in: Applications of ion beams to materials. Carter, G., Colligon, J. S. Grant, W. A. (eds.). London: Inst. Physics 1976, p. 154
237. Trillat, J. J., Hayman, P., in: Le Bombardment Ionique, CNRS (1961)
238. Mezey, G., Nagy, T., Gyulai, J., Kotai, E., Lohner, T., Manuaba, A.: Proc. SPIG Conf., Dubrovnik, 27. Aug. 1976, p. 224
239. Fritzsche, G. R.: J. Electrochemic. Soc. *120*, 1603 (1973)
240. Nomura, K., Hirose, Y., in: Ion implantation in semiconductors. Namba, S. (ed.). New York: Plenum Press 1975, p. 681
241. Williams, J. S., Grant, W. A., in: Applications of ion beams to materials. Carter, G., Colligon, J. S., Grant, W. A. (eds.). London: Inst. Physics 1976, p. 31
242. Grant, W. A., Williams, J. S., Christodoulides, C. E., in: Physics of ionized gases. Navincek, B. (ed.). Ljubljana: J. Stefan Institute 1976, p. 341
243. Mackintosh, W. D., Brown, F.: Proc. Int. Conf. Application of Ion Beams to Metals. Picraux, S. T., Eer Nisse, E. P., Vook, F. L. (eds.). New York: Plenum Press 1974, p. 111
244. Towler, C., Collins, R. A., Dearnaley, G.: J. Vac. Sci. Technol. *12*, 520 (1974)
245. Crowder, B. L., Tan, S. I.: IBM Techn. Disclosure Bull. 14 (1971)
246. Spitznagel, J. A., Fleischer, L. R., Choyke, W. J., in: Applications of ion beams to metals. Picraux, S. T., EerNisse, E. P., Vook, V. L. (eds.). New York: Plenum Press 1974, p. 87
247. Ashworth, V., Grant, W. A., Procter, R. P. M., Wellington, T. C.: Corros. Sci. *16*, 393 (1976)
248. Antill, J. E., Bennet, M. J., Dearnaley, G., Fern, F. H., Goode, P. D., Turner, J. F.: Proc. Int. Conf. Ion Implantation in Semiconductors and other Materials, Yorktown Heights. New York: Plenum Press 1972
249. Antill, J. E., Bennet, M. J., Carney, R. F. A., Dearnaley, G., Fern, F. H., Goode, P. D., Myatt, B. L., Turner, J. F., Warburton, J. B.: Corros. Sci. *16*, 729 (1976)
250. Benjamin, J. D., Dearnaley, G., in: Applications of ion beams to materials. Carter, G., Colligon, J. S., Grant, W. A. (eds.). London: Inst. Physics 1976, p. 141
251. Dearnaley, G., Benjamin, J. D., Miller, W. S., Weidmann, L.: Corros. Sci., *16*, 717 (1976)
252. Dearnaley, G., in: Applications of ion beams to metals. Picraux, S. T., EerNisse, P. E., Vook, F. L. (eds.). New York: Plenum Press 1974, p. 63
253. Muhl, S., Collins, R. A., Dearnaley, G., in: Applications of ion beams to materials. Carter, G., Colligon, J. S., Grant, W. A. (eds.). London: Inst. Physics 1976, p. 147
254. Ashworth, V., Baxter, D., Grant, W. A., Procter, R. P. M.: Corr. Sci. *16*, 775 (1976)
255. Ashworth, V., Baxter, D., Grant, W. A., Procter, R. P. M., Wellington, T. C.: Ion implantation in semiconductors. Namba, S. (ed.). New York: Plenum Press 1975, p. 367

256. Ashworth, V., Baxter, D., Grant, W. A., Procter, R. P. M.: Corr. Sci. *17*, 947 (1977)
257. Ferber, H., Kasten, H., Wolf, G. K., Lorenz, W. J., Schweickart, H., Folger, H.: Conf. Ion Implantation and Ion Beam Analysis Techniques in Corrosion, 28–30 June 1978, Manchester
258. Ferber, H., Wolf, G. K.: Conf. Ion Implantation and Ion Beam Analysis Techniques in Corrosion, 28–30 June 1978, Manchester
259. Harker, G. J.: J. Soc. Chem. Ind. (London) *51*, 314 T (1932)
260. Götzky, S., Günther, P.: Z. Physik. Chem. *B26*, 386 (1934)
261. Pissarjerosky, L., Tschrelaschwili, S., Ssatotschenko, G.: Acta Physicochim. URSS, *7*, 289 (1937)
262. Weisz, P. B., Swegler, E. W.: J. Chem. Phys. *23*, 1507 (1955)
263. Spilners, A., Smoluchowski, R., in: Reactivity of solids. deBoer, J. H. (ed.). Amsterdam: Elsevier 1961, p.475
264. Schwab, G. M., Sizmann, R., Todo, V.: Z. Naturforsch. *16a*, 985 (1961)
265. Bragg, R. H., Moritz, F. L., Holtzmann, R., Feng, P. Y., Armour, F.: Res. Found., Illinois Inst. Technol. WADC Techn. Rept. 59–286 (1959)
266. Taylor, E. H.: Adv. Catalysis *18*, 111 (1968)
267. Pinel, J., Levaillant, C.: German Pat. 2218153 (1972)
268. Balarin, M.: DDR Pat. 77961 (1970)
269. Farnsworth, H. E., Woodcock, R. F.: Adv. Catalysis *11*, 123 (1957)
270. Yamashina, V., Farnsworth, H. E.: Ind. Eng. Chem. Prod. Res. Develop. *2*, 34 (1963)
271. Shooter, D., Fansworth, H. E.: J. Phys. Chem. *66*, 222 (1962)
272. Tuul, J., Farnsworth, H. E.: J. Amer. Chem. Soc. *83*, 2247 (1961)
273. Sosnovsky, H. M. C.: J. Phys. Chem. Solids *10*, 304 (1959)
274. Sosnovsky, H. M. C., Ogilvie, G. J., Gillam, E.: Nature *182*, 523 (1958)
275. Baumgärtner, F. et al.: KFK-Report Karlsruhe 2275, 171 (1975)
276. Baumgärtner, F.: KFK-Report Karlsruhe 1950, 193 (1973)
277. Bockris, J. O. M., Reddy, A. K. N., in: Modern electrochemistry. New York: Plenum/Rosetta 1970, p.1141
278. Voinov, M., Buhler, D., Tannenberger, H.: Proc. Electrochem. Soc. San Francisco (1974)
279. Hayes, M., Kuhn, A. T., Grant, W. A., cited by Grant, W. A., in: Physics of ionized gases. Navincek, B. (ed.). Ljubljana: J. Stefan Institute 1976, p.334
280. Kuhn, A. T., Wright, P. M.: J. Electroanal. Chem. *38*, 291 (1972)
281. Grennes, M., Thompson, M. W., Cahn, R. W.: J. Appl. Electrochem. *4*, 211 (1974)
282. Hartley, N. E. W., in: Applications of ion beams to materials. Carter, G., Colligon, J. S., Grant, W. A. (eds.). London: Inst. Physics 1976, p.210
283. Hartley, N. E. W.: Tribology *8*, 65 (1975)
284. Pavlov, P. V., Zorin, E. I., Tetelbaum, D. I., Lesnikov, V. P., Ryzhkov, G. M., Pavlov, A. V.: Phys. Stat. Solidi *19*, 373 (1973)
285. Kanaja, K., Koya, K., Toki, K.: J. Phys. E: Sci. Instrum. *5*, 541 (1972)
286. Hartley, N. E. W.: J. Vac. Sci. Technol. *12*, 485 (1975)
287. Gabovich, M. D., Budernaga, L. D., Poritskii, V. Y., Protsenko, J. M.: Proc. All Soviet Meeting on Ion Beam Physics, Kiev (1974)
288. Hartley, N. E. W., Dearnaley, G., Turner, J. F., Saunders, J.: Proc. Int. Conf. on Applications of Ion Beams to Metals. Picraux, S. T., EerNisse, E. P., Vook, F. L. (eds.). New York: Plenum Press 1974, p.123
289. Hartley, N. E. W.: Wear *34*, 427 (1975)

Received October 9, 1978

Doubly-Charged Negative Ions in the Gas Phase

Robert W. Kiser

Department of Chemistry, University of Kentucky, Lexington, Kentucky 40506, U. S. A.

Gas phase electron impact processes give rise to both positive and negative ions, although the negative ions commonly constitute ca. 0.1 % of the total ions formed. Lattice energy calculations which consider the existence of doubly-charged negative ions (such as O^{2-} and S^{2-}) in the crystalline state, indicate that the process

$$Z^- \ (g) + e \ (g) \rightarrow Z^{2-} \ (g)$$

is endoergic by about 6–8 eV. Here, Z is any element in general. However, doubly-charged negative ions are now known to exist in the gas phase. Their existence and properties are of interest to many scientists; this review assesses the current information concerning these unique and important species.

The lifetimes of gaseous doubly-charged negative ions in their ground state are expected to be short and have been predicted to be of the order of 0.1 ns or less, a time so short as to preclude their observation in a conventional mass spectrometer. However, X^{2-} (g) is isoelectronic with G^- (g), where X is a halogen and G is a noble gas element; the He^- (g) species has been known since 1939, has been studied by several investigators who have determined the lifetimes of He^- (g) [$1s^1 2s^1 2p^1 \ ^4P$], and has been utilized as a projectile. It is shown that the X^{2-} species also exist generally in excited states, not unlike the He^- ion, and that they have long lifetimes of approximately 0.1 ms.

The discovery of the existence of Z^{2-} (g) and AZ^{2-} (g) ions was reported in 1966 in experiments that employed an omegatron mass spectrometer. The experiments involved in the discovery of doubly-charged negative ions are reviewed, as are those of other investigators who have reported observations of atomic Z^{2-} (g) ions and molecular AZ^{2-} (g) ions. Mechanisms and methods of formation of the doubly-charged negative ions and their energetics and lifetimes are discussed in an effort to improve the understanding of this new class of gaseous ions.

R. W. Kiser

Table of Contents

I Introduction

A dozen years ago experimental evidence for the existence of doubly-charged negative ions in the gas phase was first reported by Stuckey and Kiser[280, 281]. Subsequent observations and reports by other investigators have confirmed the existence of these novel ions. It has been established that these unique species can have lifetimes of the order of 0.1 ms or longer. Still, little is known of the mechanisms of formation and of the properties of these doubly-charged negative ions.

The present paper reviews and assesses the current information about the formation and identification of X^{2-}, AZ^{2-}, and other doubly-charged negative ion species that have been reported. It also presents mechanisms and explanations for the production of the monatomic, diatomic and polyatomic doubly-charged negative ions that have been observed. Finally, some potentially fruitful paths for the further study of these ions are indicated.

To the present, experimental evidence has been presented for the gas-phase existence of the monatomic species H^{2-}, F^{2-}, Cl^{2-}, Br^{2-}, I^{2-}, O^{2-}, Te^{2-}, P^{2-}, As^{2-}, Sb^{2-}, and Bi^{2-}, for the diatomic CN^{2-} ion, and for the polyatomic species NO_2^{2-}, SF_5^{2-}, SF_6^{2-} and others. Before examining the experimental studies of these doubly-charged negative ions in the gas phase, it is instructive to consider information concerning species isoelectronic with doubly-charged negative ions.

A. Isoelectronic Species of X^{2-} Ions

Species isoelectronic with the doubly-charged halogen ions, X^{2-}, e. g., Cl^{2-}, will be considered first. As part of a preliminary communication concerning an investigation of singly- and doubly-charged positive ions formed by electron impact with argon, Barton[17] noted that large quantities of negative ions also were produced. He surmised that these negative ions were Ar^- which apparently originated in a region of the discharge tube where impacting electrons had just sufficient energy to excite the atoms. Barton[17] commented that in argon it appears to be a necessary and sufficient condition for the capture of an electron that the atom first be in an excited state. However, in his full paper[18] published on Ar^+ and Ar^{2+} formation, no comment is made concerning these negative ions. A separate contribution[19] later that year does treat these negative ions and Barton revised the assignment of these ions as due to Ar^-; rather, he demonstrated that the negative ions hat M/z = ca. 33 and were due to the oxide-coated filaments he employed. Consequently, he assigned the species he observed as O_2^- rather than Ar^-.

The existence of He^- was first estabished by Hiby[140] in 1939 when he observed H_2^- and He^- in canal rays from hydrogen and helium with a parabola mass spectograph[287]. Although Döpel[93] had reported the He^- species earlier, Hiby[140] gives reasons that indicate Döpel's interpretation was based on an illusion and was dubious. Also using a parabola mass spectrograph, Tüxen[293] did not find He^- to be present in a helium glow discharge over the pressure range of 0.1 to 10 Torr; in fact, he found no negative ions of the noble gases to be formed from He, Ne or Ar. However, Hiby[140] found that the He$^-$ ion intensity was about 1 % of that of the He$^+$ ion produced from pure helium.

The He$^-$ ion has since been extensively studied both theoretically and experimentally. Bartlett[16] compared his extrapolation results for H with those of Bethe[32] and concluded that $1s^2 2s^1$ ^{1}S He$^-$ is unstable. Wu's variational calculations[314] indicated He$^-$ in the $1s^2 2s^1$ ^{2}S state to be unstable. However, he found that the $1s^1 2s^2$ ^{2}S state of He$^-$ is very close to the energy of the $1s^1 2s^1$ ^{3}S state of He and indicated that this excited state of He$^-$ might just be stable. In a similar manner he calculated Li in the $1s^2 2s^1$ ^{2}S state to have an electron affinity of 0.54 eV, a value close to the currently accepted experimental value. Wu and Shen[315] indicated that the He$^-$ ion in the $1s^1 2s^1 2p^1$ ^{4}P state would be possible.

Holøien[150] considered various excited states of He$^-$ using the variational method and showed that the $1s^1 2p^2$ ^{4}P state is unstable and subject to autodetachment. He also found the bound $2s^1 2p^2$ ^{4}P state to be subject to autodetachment. Since permitted autodetachment would cause the lifetimes to be of the order of 10^{-14}s, neither of these states could explain the lifetime of $> 10^{-7}$ s necessary for He$^-$ to be detected mass spectrometrically. Holøien[150] noted that Wu and Shen[315] suggested the $1s^1 2s^1 2p^1$ ^{4}P state might be stable, and indicated that this ^{4}P state would not undergo autodetachment.

Subsequently, Holøien and Midtdal[152] treated the $1s^1 2s^1 2p^1$ ^{4}P state of He$^-$. In their detailed theoretical treatment, Holøien and Midtdal found that the only stable electronic configuration of the He$^-$ ion is $1s^1 2s^1 2p^1$ ^{4}P and that this state is metastable toward radiative decay and is not subject to autodetachment. From the energy calculations they found an electron affinity for $1s^1 2s^1$ ^{3}S He relative to this state of He$^-$ of at least 0.075 eV, although a later correction of a calculational error causes this value to become negative[151]. The original value was quite close to the experimental photodetachment measurements of Brehm, Gusinov and Hall[50]. Holøien and Geltman[151] have calculated an electron affinity of > 0.033 eV; these authors also cite an unpublished calculation of Weiss that yielded an electron affinity of > 0.069 eV. Holøien and Midtdal[152] note that this ^{4}P$_{5/2}$ state of He$^-$ might exist for times of the order of 1 ms and that this then explains the observations of Hiby[140]. Pietenpol[239] calculated the lifetime of the ^{4}P$_{5/2}$ state of He$^-$ to be 1.7 ms and Laughlin and Stewart [176] have calculated that this He$^-$ ion would have a lifetime of about 0.4 ms. Manson[191] calculated a lifetime of about 1 ms for the ^{4}P$_{5/2}$ state and Estberg and LaBahn[105] have calculated a lifetime of about 0.45 ms for this ^{4}P He$^-$* ion.

Nicholas, Trowbridge and Allen[223] experimentally produced He$^-$ by charge exchange from He$^+$ formed in a cold-cathode mercury pool discharge and measured the lifetime of the He$^-$ beam to be about 0.018 ms. They believed that they were studying a single type of He$^-$ ion and assumed it to be the $1s^1 2s^1 2p^1$ state. Earlier, Riviere and Sweetman[250] obtained evidence for a finite lifetime for the He$^-$ ion and Sweetman[283] experimentally determined the lifetime of He$^-$ to be > 1.0 x 10^{-5} s. Based upon the electric field dissociation results reported by Riviere and Sweetman[250] and the theoretical treatment by Demkov and Drukarev[85] of the decay of negative ions in an electric field, it is estimated that the He electron affinity is very near 0.08 eV. Smirnov[268] has discussed the decay of negative ions and

Smirnov and Chibisov[270] determined an electron affinity for He of 0.06±0.005 eV. More recently, Blau, Novick and Weinflash[39, 224] and Simpson, Browning and Gilbody[266] have determined the $^4P_{5/2}$ He⁻* lifetime to be about 0.35–0.50 ms; they also identified a shorter-lived component with a lifetime of about 0.10 ms that corresponds to the other ^4P states, in agreement with the theoretical calculations. Using an rf resonance technique, Mader and Novick[189, 190] have studied the fine structure of the $1s^1 2s^1 2p^1$ ^4P_J He⁻ ion.

Purser[243] has suggested a mechanism for the production of He⁻. The first step is the formation of the excited ^3S He atom. In the second step, electron attachment to the ^3S He produces the bound $1s^1 2s^1 2p^1$ ^4P state of He⁻.

The He⁻ ion and other negative ions are of practical importance in high energy nuclear physics. Using a tandem accelerator, negative ions are first accelerated and then, by means of charge-stripping with a thin target, are converted to positive ions for further acceleration[104, 159, 251]. Windham, Joseph and Weinman[307] have described a He⁻ ion source that produces He⁺ in a discharge. The He⁺ ions then are caused to undergo collisions with hydrogen to form He⁻ (and H⁻) by a charge exchange process. In this way, yields of > 3 nA of He⁻ were produced. Negative ion charge exchange has been studied by Tang, Rothe and Reck[284] using Cs bombardment of SO_2, SF_6, CF_2Cl_2 and CH_3NO_2. A number of atomic and molecular negative ions so produced have been investigated (Z^{2-} ions were not reported, however).

Recently, Muller, Lorentz and Remy[216] used a duoplasmatron ion source for the production of He⁻ ion beams. Interestingly, their source design permits partial magnetic separation of electrons and weakly bound negative ions in the extraction zone. These same authors[44] have also produced H⁻, F⁻ and Cl⁻ using their duoplasmatron; they note that very low energy electrons are present in the anodic plasma where Cl⁻ and F⁻ are formed from CF_2Cl_2. Other negative ion sources for tandem accelerators have been surveyed[80, 209]. Purser[243] also has reviewed the advances in negative ion source technology and notes that beam currents in the μA to mA range are now being produced for some negative ion species.

In summary, the He⁻ ion has been produced, studied and employed experimentally. Theoretical calculations indicate that this ion is the $1s^1 2s^1 2p^1$ $^4P_{5/2}$ state, a state metastable with respect to both radiative decay and autodetachment. The results of theoretical calculations and experimental measurements are in reasonable agreement that the electron affinity of $1s^1 2s^1$ ^3S He to form He⁻ is 0.08 ± 0.02 eV and that the lifetime of this metastable He⁻ ion is about 0.45 ms. Resonant states of He⁻ are also known and have been characterized[106, 222].

Thus, nearly forty years after the discovery of He⁻ by Hiby, this interesting ion is beginning to be understood. The other singly-charged noble gas negative ions, G⁻, are less well understood, although progress is being made. Quéméner, Paquet and Marmet[244], Bolduc, Quéméner and Marmet[43] and Marmet, Bolduc and Quéméner[192] have studied resonant negative ion states of He (e. g., $2s^2 2p^1$ ^2P and $2s^1 2p^2$ ^2D), Ar ($3s^1 3p^6 4s^2$), Kr, and Xe by electron impact. Zollweg[318] has indicated none of the rare gases have bound states. Moser and Nesbet[215] have used variational Bethe-Goldstone calculations to compute the electron affinity of Ar (and other atomic species) and Edwards[102] has studied the doubly-excited ^2P states of

Ar^-. Long-lived metastable states of negative ions of the noble gas atoms other than He are unknown to the present, although Bethge[34] indicated he has extracted very low intensity Xe^- ion beams from his dense plasma Penning source. Since so little other information is available at this time, it is not inappropriate to adopt He^{-}* as a model for the isoelectronic monatomic X^{2-} ions of Group VII and for other Z^{2-} species.

Holoien and Geltman[151] also have found that the $1s^1 2p^2$ $^4P^e$ state of He^- is bound with respect to the $1s^1 2p^1$ $^3P^0$ state of He; the electron affinity of this state is calculated to be $\geqslant 0.20$ eV. This state is metastable against autodetachment via the Coulomb interaction. Holoien and Geltman indicate its lifetime against autodetachment to be of the same order as that for the $^4P^0$ state and that it may radiate to the metastable $^4P^0$ state. The extent of contributions of the $^4P^0$ and $^4P^e$ metastable states to the observed He^- ion beams is presently unknown. Buchel'nikova[57], Smirnov[269] and Massey[196] have summarized much of the present knowledge of atomic negative ions.

Other species isoelectronic with He^-, e. g., Li^-[109] have been treated in detail by many of the same investigators who have examined He^-. A particularly important species that is isoelectronic with He^- and is of interest to this discussion is H^{2-} The H^- ion is known to have at least one bound state[134, 173, 182, 233, 253, 254] (a very extensive literature exists concerning the H^- ion). Recently, Hills[142] has reported that the singlet state of H^- has only one bound state and that the triplet state of H^- has no bound state within the approximations he employed. Further comments on the doubly-charged negative H^{2-} ion, however, will be held is abeyance until the experimental observations of a long-lived H^{2-} ion by Schnitzer and Anbar[255] and a much shorter-lived H^{2-} ion by Walton, Peart and Dolder[303] are considered.

Doubly-charged negative ions of Group VI elements are isoelectronic with the halogen ions, X^-, and with the noble gases, G. Within the past 15 years, the electron affinities of the halogen atoms and numerous other atoms have been determined precisely by means of photodetachment[48]. However, there is a paucity of information about any excited states of the halogen ions. Most of the available information has been derived from isoelectronic sequence extrapolations and, as such, is less than conclusive. In view of the lack of other information, this approach will necessarily be considered in the discussion of the Z^{2-} species. Z is used as a general designation for any element. What is to be noted for the present is that the calculation of electron affinities with closed shell configurations for Z^{2-} yields negative values, indicating that the Coulombic repulsion is too great to achieve a bound state. For example, Herrick and Stillinger[136] indicate the electron affinity of O^- is -5.38 eV based on a variational treatment. This necessitates the consideration of excited states that will be metastable both toward radiative decay and autodetachment, although Avron, Herbst and Simon[11] have indicated that normally unbound states may be weakly bound in magnetic fields.

A few Z^{2-} species are rather well characterized in the solid state. It is then appropriate to consider the information about Z^{2-} (s) with the intent of learning more about Z^{2-} (g).

B. Doubly-Charged Negative Atomic Ions in Solids

Several solid compounds of a predominantly ionic nature have doubly-charged negative monatomic ions as the anions. Foremost among these are the oxides of the strongly electropositive elements of Groups I and II. Relatively fewer of these elements will form compounds with sulfur or selenium with more than 50 % ionic character[230, 231].

Pritchard[242], Buchel'nikova[54, 57], Branscomb[48], Moiseiwitsch[211], Berry[30], Berry and Reimann[31], Christophorou[61], Franklin and Harland[112], Hotop and Lineberger[153], Moiseiwitsch[212], Schoen[257] and Steiner[278] have reviewed the theoretical treatments and experimental measurements of the electron affinities of many species. Negative ions and negative ion mass spectra have been discussed also by Buchel'nikova[56], Khvostenko and Tolstikov[165], Massey[195−197], Mazunov and Khvostenko[200], McDaniel[201], Melton[206, 207], Page and Goode[225], Schulz[259] and von Ardenne, Steinfelder and Tümmler[297].

Table 1. Electron affinities of selected atomic and molecular species (in eV). [Estimated values for O^- and S^- are in brackets.]

Species	Electron affinity
CN	3.82
SF_5	3.66
Cl	3.61
Br	3.36
F	3.34
I	3.06
CO_3	2.69
BF_3	2.65
NO_2	2.36
O_3	2.26
S	2.08
Se	2.02
Te	1.97
Bi	1.76
O	1.46
SF_6	1.43
C	1.27
H	0.76
P	0.75
Li	0.61
Na	0.54
K	0.50
Rb	0.49
Cs	0.47
He ($1s^1 2s^1\ ^3S$)	0.08
S^-	[−6.0]
O^-	[−7.7]

Table 1 presents a list of the electron affinities for several selected atomic and molecular species of interest to the present discussions; results determined by the photodetachment technique have been selected from the literature. No experimental measurements of the electron affinity of O^- or S^- are available. However, lattice energy calculations[58, 175, 242] and a variety of isoelectronic extrapolation techniques have been employed by a number of investigators to obtain this information. Baughan[22, 23], for example, finds values of -6.2 and -3.9 eV for the attachment of two electrons to the O and S atoms, respectively. These results provide the estimates indicated in Table 1 for $EA(O^-)$ and $EA(S^-)$.

It is recognized that a calculated negative electron affinity does not mean that the negative ion cannot exist. Species such as Be^-, Mg^-, BeH^-, MgH^-, LiH^- and LiH_2^- have been formed in a Penning ion source[25, 36]. These species are formed in excited states, and metastable states provide long lifetimes against autodetachment and radiative decay. Because the He^- ion was not observed, but Be^-, BeH^-, Mg^- and MgH^- were, Bethge, Heinicke and Baumann[36] indicate that the lifetimes of these ions must be greater than 10^{-5} s, the lifetime experimentally determined for He^{-*} by Sweetman[283] and by Nicholas, Trowbridge and Allen[223]. Considering the more recent determinations of the lifetime of He^{-*} [39, 224, 266], the Be^- and Mg^- ions and the corresponding hydride species must have lifetimes of the order of 10^{-4} s or greater. Weiss[306] has suggested that the Be^- and Mg^- ions observed by Bethge, et al.[36] may be the metastable $ns^1 np^2\ ^4P$ states that have positive electron affinities.

The inference is that the metastable states of the negative ions and the long lifetimes of these states must play significant roles in the formation of doubly-charged negative ions in the gas phase. In the solid state the crystal lattice energy apparently stabilizes the Z^{2-} species. As mentioned, $O^{2-}(s)$, $S^{2-}(s)$ and $Se^{2-}(s)$ are known in many compounds. But $Cl^{2-}(s)$ and other doubly-charged negative halogen ions in the solid state are unknown. For a metal M exhibiting a valence of n+ in an ionic compound that might contain the doubly-charged negative halogen ion, X^{2-}, a general empirical formula of $M_2 X_n$ would result. It is of interest to consider if any such $M_2 X_n$ compounds are known to exist.

A number of silver subhalides, of inconstant composition and apparently colloidal solutions of the metal in the halide, have been reported. These are not true compounds. The literature[263] does record an interesting species known as silver subfluoride. Parkes[226] indicates this $Ag_2 F$ to be a relatively unstable bright yellow crystalline solid prepared by electrolysis of an AgF solution[169, 230]. Wöhler[312] characterized the crystals not as yellow or bronze, as did Guntz[122], but rather as light green to bronze metallic. The $Ag_2 F$ crystals can also be prepared by heating an HF solution of AgF with silver powder. The compound is unstable in contact with water if there is no excess of HF, and decomposes above 90 ° to silver and silver fluoride. Thorne and Roberts[288] comment further on the properties of $Ag_2 F$. One may question if this compound is best represented as $2Ag^+$, F^{2-}. The apparent answer is that there is no reason to believe X^{2-} ions are present in this crystalline solid or any other reported $M_2 X_n$ substance.

If it is assumed that $Ag_2 F$ is composed of Ag^+ and F^{2-} ions, a crude estimate of the electron affinity of F^- may be made by employing a Born-Haber thermochemical cycle. The assumptions made are that ΔHf (Ag_2F, rutile structure) is ca. -0.8 eV and that the ionic radius of an F^{2-} ion is ca. 50 % larger than F^-, or ca. 2 Å. An approximate crystal lattice energy for Ag_2F may be calculated as -17.7 eV using the Born-Mayer equation. EA (F), $D(F_2)$, S(Ag) and I(Ag) are well known values[301]. Therefore, an estimate of the energy associated with the attachment of another electron to the F^- ion is that $EA(F^-) = 1.7$ eV. A larger ionic radius for F^{2-} would cause $EA(F^-)$ to be greater; a more positive $\Delta Hf(Ag_2F)$ would cause $EA(F^-)$ to be less.

The extrapolation of ionization potentials along the isoelectronic sequence of Na, Mg^+ and Al^{2+} by a variety of methods all indicate that F^{2-} is not bound in the ground state (i.e., $EA(F^-)$ is ca. -7 eV[280]). The disparity of this result with the result from the thermochemical calculation indicates that Ag_2F is not composed of Ag^+ and F^{2-} ions. Pauling[230] has indicated that, based on X-ray structure studies[139, 279], silver subfluoride may be considered to be intermediate between a salt and a metal. The crystal is composed of Ag,Ag,F layers with the bonds between silver layers being essentially metallic and those between silver and fluorine layers being ionic-covalent in nature. Consequently, this author is unaware of any principally ionic compounds that possess the general empirical formula M_2X_n that contain X^{2-} ions. There appears to be no known compounds that contain halogen X^{2-} ions in the solid state.

The fact that the ionic model works rather well for crystal lattice energy calculations of compounds such as MgO and Li_2O cannot be interpreted as establishing that O^{2-} actually exists in the solid. Crystal lattice energy calculations of Mg_3N_2, Al_2O_3, and AlN are in reasonable accord with experiment, but certainly all of these compounds have significant covalent character. Crystal lattice energy calculations based on the ionic model are found to be relatively insensitive to a significant covalent contribution to the bonding. Ahlrichs[3] makes a similar comment using MgO as an example in his discussion of the Hartree-Fock theory for negative ions.

Driessler, Ahlrichs, Staemmler and Kutzelnigg[95] have discussed some of the difficulties encountered in Hartree-Fock computations of polyatomic negative ions. Ahlrichs[3] indicates that for the O^{2-} system in the ^1S_g state, the maximum number of electronic charges that can be on the O atom is z = 1.23; similarly, for the ^1S_g S^{2-} system, z = 1.30. Ahlrichs[3], Prat[241] and Delgado-Barrio and Prat[84] do not treat the various possible excited states of O^{2-} (analogous to those discussed in detail above for the He^- system) in their Hartree-Fock calculations. Herrick and Stillinger[136] also have treated the ground state of O^{2-} using a variational approach.

With this background based on the isoelectronic sequence species and electron affinities, and the electronic configurations of various negative ions, a review of the discovery of gas phase doubly-charged negative ions and an examination of the various experimental approaches and observations of doubly-charged negative ions by a number of different workers is appropriate.

The literature[107, 195, 201, 220] records a variety of statements to the effect that long-lived ($> 10^{-7}$ s) doubly-charged negative ions do not exist. The remainder of this discussion basically involves descriptions and evaluations of experiments and interpretations that demonstrate that doubly-charged negative ions do exist.

II The Discovery of Doubly-Charged Negative Ions

The discovery of doubly-charged negative ions was first reported in 1966 by Stuckey and Kiser[280, 281]. The species observed were the monatomic O^{2-}, F^{2-}, Cl^{2-}, and Br^{2-}, and the diatomic CN^{2-} ions. Their studies also included an estimate of the lifetime of these species based on the ion flight times and a discussion of the electronic configurations and possible modes of formation of the observed ions.

The investigations by Stuckey and Kiser employed a rather unique type of mass spectrometer which, as will be seen, is important in the production of the doubly-charged negative ions. We shall review briefly the experimental approaches, observations, and initial interpretations of the results. To do so, an examination of the design and characteristics of the omegatron is germane.

A. The Omegatron Mass Spectrometer

Hipple, Sommer and Thomas[143] devised a mass spectrometer of new design to determine the Faraday by measuring the cyclotron resonance frequency of protons in a known magnetic field. Recognizing that this device measures ω, the cyclotron frequency, Hipple, Sommer and Thomas proposed the name *omegatron*.

The basic design and operation of this mass spectrometer are closely related to that of the cyclotron devised by Lawrence and Livingston[177]. De Groot[81] has also discussed the cyclotron operation. (Interestingly, ^3He and ^3H were found by Alvarez and Cornog[8, 9] using the cyclotron as a mass spectrometer.) A detailed discussion of the omegatron was given by Hipple, Sommer and Thomas in subsequent publica-

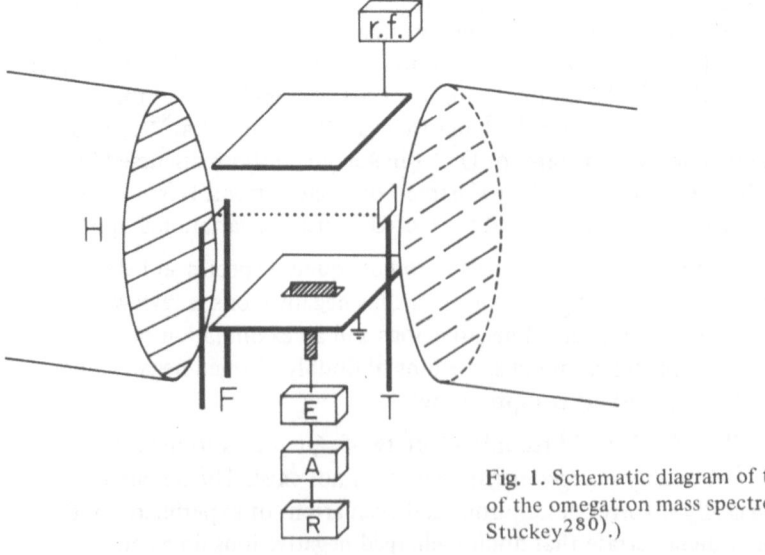

Fig. 1. Schematic diagram of the basic design of the omegatron mass spectrometer. (After Stuckey[280].)

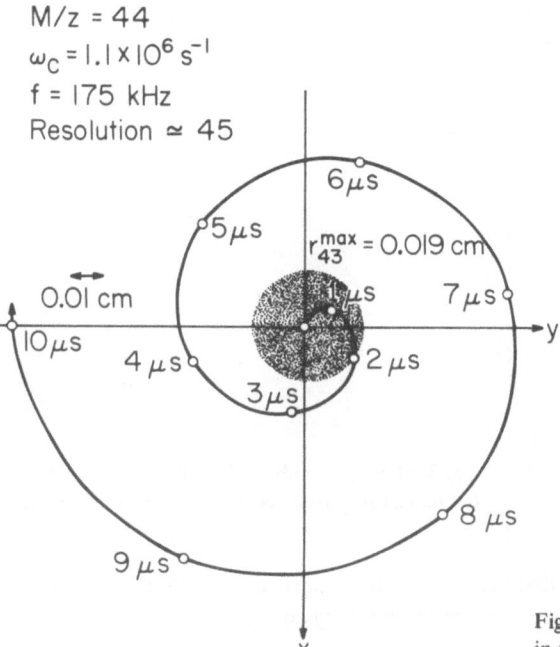

M/z = 44
$\omega_c = 1.1 \times 10^6 \text{ s}^{-1}$
f = 175 kHz
Resolution ≈ 45

Fig. 2. Ion trajectory of a resonant ion in the omegatron mass spectrometer

tions[144, 273]; the presentation here is based essentially on their reports and the detailed ion trajectory analyses provided by Berry[28, 29], Brubaker and Perkins[52], Schuchhardt[258] and Truell[291, 292]. Other treatments and discussions are also available[108, 167].

The design of the omegatron is shown schematically in Fig. 1. The omegatron functions as a small cyclotron with the important distinction that the radio-frequency electric field is distributed over the entire analyzing region and is not just applied across the faces of two "D"s as in the cyclotron[81, 177].

An ion of mass m and charge q, the resonant frequency of which is equal to the applied radio-frequency, f,

$$f = qB/2\pi m = \omega_c/2\pi \tag{1}$$

continuously gains energy from the rf field; as a result, the radius of the ion trajectory continues to increase until the ion strikes the target electrode and is detected, as shown in Fig. 2. This ion trajectory closely approximates an Archimedes spiral. ω_c is the cyclotron frequency and B is the magnetic field. The radius at any given time, t, is given by

$$r = E_o t/2B \tag{2}$$

E_o is the rf field strength.

Ions out of phase with the radio-frequency field cannot gain energy continuously from the field; as a result, these ions do not exceed a radius, r (max), that is signif-

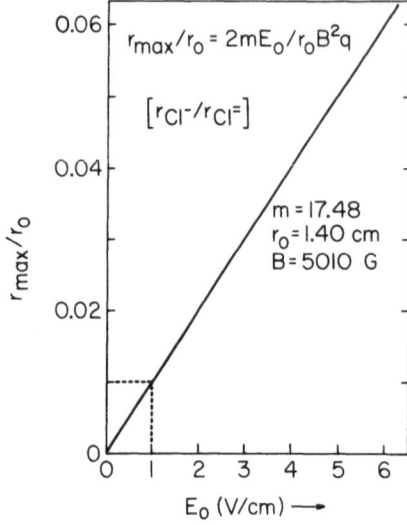

Fig. 3. Maximum radius of a non-resonant ion in the omegatron mass spectrometer. (After Stuckey[280])

icantly less than the straight-line distance between the electron beam and the detector, r_o. The maximum radius of a non-resonant ion is given by (see also Fig. 3)

$$r(max) = (m/\Delta m) (2mE_o/qB^2) = (m/\Delta m) (r_o/\rho) \qquad (3)$$

for ions differing in mass by Δm from resonant ions of mass m. Here ρ is the resolution. Consequently, the non-resonant ions do not reach the detector. Rather, these non-resonant ions generate complex rosettes[143, 144, 273] and remain in the vicinity of the electron beam, as indicated schematically by the dotted region in Fig. 2. A distinguishing characteristic of the omegatron, therefore, is the fact that not all ions are extracted promptly from the ion source or region of the electron beam.

As seen from Eq. (3), r(max) is dependent directly upon E_o. Figure 3 graphically indicates the ratio of $r(max)/r_o$ using typical operating conditions employed by Stuckey and Kiser as an example. For a resonant Cl^{2-} ion, the maximum radius of the non-resonant Cl^- ions is about 0.015 cm at $E_o = 1.0$ V/cm when a constant magnetic field of 5010 G is used. Thus, the non-resonant Cl^- ions execute very tight rosette patterns and remain essentially with the electron beam.

Figure 2 gives an indication of the ion trajectory of a resonant ion in the first 0.01 ms in the omegatron. The flight time of the ion to the detector r_o distant from the electron beam is approximately given by

$$t = 2Br_o/E_o \qquad (4)$$

so that for $E_o = 1.0$ V/cm, a flight time of about 0.14 ms is required (see Fig. 4), and this is independent of m and q. Longer flight times can be effected by decreasing E_o. For an ion of M/z = 44 f = 175 kHz and $\omega_c = 1.1 \times 10^6$ s^{-1}, and the ion will make about 25 revolutions in a field of 5010 G to reach the detector.

The resolution of the omegatron varies directly with B^2 and r_o, and inversely with E_o and m/q. In order to maintain $r(max)/r_o$ at a small value, E_o must be held at a

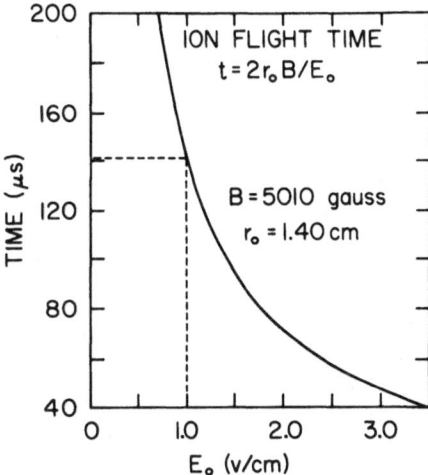

Fig. 4. Flight time of resonant ions in the omegatron mass spectrometer. (After Stuckey[280].)

low value. Lower rf field strengths permit less energy to be derived from the rf field per revolution, so that a longer path length and better resolution result. However, the sensitivity also depends upon E_o. Thus, for the experiments performed by Stuckey and Kiser with an omegatron that was essentially a General Electric ion resonance spectrometer[214], the selection of $E_o = 1.0$ V/cm provided good sensitivity and reasonable resolution. B and r_o were fixed by the experimental apparatus used[280]. The resolution varied inversely with m/q and was about 45 at M/z = 45 (i.e., $\rho = 2000$ at M/z = 1). This resolution was adequate for the study of the two isotopic species of Br^{2-} (M/z = 39.5 and 40.5).

The arrangement of ion-trapping plates used in the omegatron for the selection of positive or negative ions is shown in Fig. 5. By use of a trapping potential the ions are held in a potential well. Negative ions are selected for study by the application of a suitable trapping voltage. The dc potential is applied to the plates in a manner closely related to that used by Sommer, Thomas and Hipple[273]. By applying a negative dc voltage to two plates symmetrically placed on either side of the electron beam, not only are the negative ions held in the plane of the detector to increase sensitivity but also positive ions can be removed from the ion beam. This trapping technique has been employed to reduce axial loss of ions in several omegatron designs[52, 273, 311, 317]. The resonant ions are then recorded as a result of neutralization at the detector; non-resonant ions held in the vicinity of the electron beam only slowly drift to the trapping plates for removal from the cool plasma.

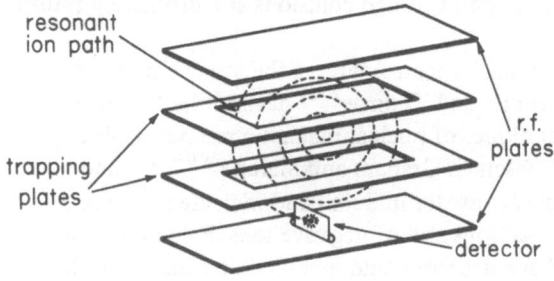

Fig. 5. Selection of positive or negative ions by means of a trapping potential in the omegatron mass spectrometer. (After Stuckey[280].)

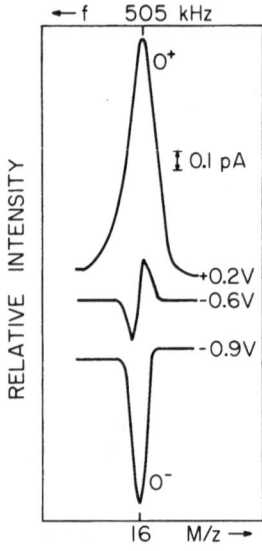

Fig. 6. Effect of trapping potential on the detection of O^+ and O^- ions formed from carbon monoxide in the omegatron mass spectrometer. (After Stuckey[280].)

The trapping voltage has essentially no effect on the resolution. However, different optimum voltages exist for positive and negative ions. For positive ions, the optimum was found by Stuckey and Kiser to be about +0.2 to 0.3 V, whereas for negative ions it was found to be about −1.0 to −1.5 V. The effect of the trapping potential on the ions extracted and detected is shown in Fig. 6 for O^+ and O^- ions formed from carbon monoxide. Stuckey[280] also noted that there is a slight frequency shift between otherwise nearly identical positive and negative ions. Similar findings also have been reported by Llewellyn[183]. This shift has been attributed to space charge effects.

Stuckey[280] has pointed out that metastable transitions would not be observed in the omegatron under common operating conditions. Consider an ion in a resonant ion path being spiraled outwards. The daughter ion formed as the result of a metastable transition somewhere along its trajectory will not be in resonance with the rf field. As a result, it will be unable to continue to derive energy from the rf field so as to reach the detector. It will continue in a circular orbit with a radius dependent upon (1) the kinetic energy acquired by the mother ion from the rf field prior to the transition and (2) the ratio of the m/q values of the daughter and mother ions. Since the lifetimes of many species that will undergo a metastable transition are $< 10^{-5}$ s, this radius will commonly be much less than r_0. Over a period of time, the kinetic energy of the daughter ion will be degraded through collisions and ultimately return to the vicinity of the electron beam.

Closely related to the omegatron mass spectrometer is the ion cyclotron resonance (ICR) mass spectrometer. Sommer and Thomas[272] devised the ICR mass spectrometer by combining the techniques of nuclear magnetic resonance absorption[40] and the basic omegatron. Wobschall, Graham and Malone[309-311] and Henis[132] have described the ICR spectrometer in detail and indicated its applications to the study of ion-molecule reactions and of negative ions. Baldeschwieler and Woodgate[13, 14] have reviewed ICR spectroscopy and included discussions of the double resonance technique. The use of ICR spectroscopy has become very important in many studies of gaseous ions; several other recent publications[125, 178] are available that review ICR spectroscopy.

As noted, the omegatron and the ICR spectrometers are very similar, particularly with respect to the method of producing ions. Sommer and Thomas[272] modified the basis of ion detection from collection at an electrode to the use of ion cyclotron resonance absorption. But, as Wobschall, Graham and Malone[309, 310] point out, the main difference is that the ion production region and the measurement region are (or can be) separated in the ICR spectrometer, but not in the omegatron.

The theoretical treatment of ion trajectories in the omegatron by Berry[29] led him to the conclusion that the omegatron will not exhibit peaks due to harmonics. That is, ions other than those of an m/q dictated by the cyclotron frequency of the ion in resonance cannot continue to gain energy without limit. Unlike the cyclotron, where particles of mass m, 3m, 5m, etc., can appear at the accelerating gap at the proper time to receive energy, the particles are in the accelerating field at all times in the omegatron, and therefore no harmonic effects exist. Stuckey and Kiser[280, 281] made a special effort to search for harmonic peaks (see below) and were unable to observe any. The results of an ICR observation of Z^{2-} species (*vide infra*) essentially rule out harmonic effects.

Stuckey[280] recognized that the residence time of a non-resonant ion is sufficiently long to become an important feature of the omegatron. The formation of the doubly-charged negative ions was attributed primarily to this long residence time. Stuckey neither predicted nor measured the residence times of non-resonant ions. However, this author estimates (from experimental data[202] for similar species) a drift velocity of some 2×10^{-5} to 2×10^6 cm/s for positive ions with a negative trapping potential of -1.0 V/cm for pressures of about 3×10^{-5} Torr. Thus, ions of the opposite charge will require of the order of 10^{-6} s to drift to a trapping plate some 0.4 cm distant (on the average). Ions of the same charge as the trapping potential will remain much longer.

B. Experiments and Discussion

The size of the basic omegatron chamber used by Stuckey and Kiser was 4.62 x 3.16 x 1.57 cm; the latter dimension defined the length of the axial electron beam. Electrons were obtained thermionically by ac heating of a 0.0025 x 0.076 cm rhenium ribbon. Mass spectral measurements were made with a nominal electron energy of 70 eV at ion source pressures of the order of 10^{-5} Torr from samples introduced through a gold leak. A trap current of 2 microamperes was commonly used, with the electron trap biased 3 V positive with respect to the chamber. The ion source was operated at 250 °C and was baked at 400 °C. A General Radio type 1001A signal generator was used together with a Keithley model 610A electrometer and a permanent magnet of 5 kG.

The detector used was a 0.5 mm thick stainless steel blade that extended less than 0.1 cm through the grounded plate into the omegatron chamber. A constant speed (3 rpm) motor and gear reduction train was used to vary the vernier control on

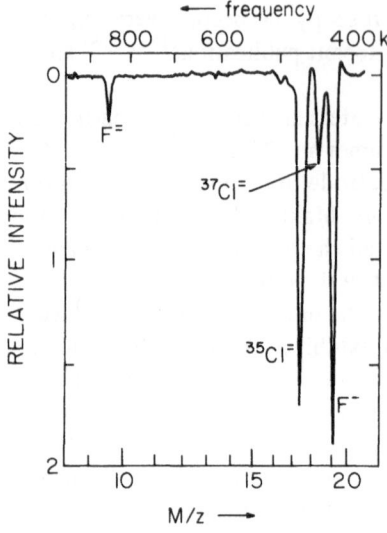

Fig. 7. Partial negative ion mass spectrum of CF_3Cl. The $^{35}Cl^-$ and $^{37}Cl^-$ peaks are not shown, but have an intensity of about ten times the $^{35}Cl^{2-}$ and $^{37}Cl^{2-}$ peaks. Electron energy, 70 V; trap current, 2 microampere; trapping potential, −0.75 V; radio-frequency voltage, 1.5 V/cm; 5 kG permanent magnet. (After Stuckey and Kiser[280, 281].)

the General Radio signal generator and thereby sweep the frequency output of the signal generator. The rf frequency was read from the panel meter of the signal generator and was monitored continuously with a Tektronix model 515 oscilloscope. The rf signal generated was an exponential function of distance (typical correlation coefficient = 0.99987) on a 10-inch 0–10 mV Brown strip chart recorder. Spectral tracings were reproduced to better than ± 2 % of the intensity in successive traces; successive traces superimposed indicated no perceptible variations in frequencies (or drive motor speed).

That doubly-charged negative ions exist is recognized from Fig. 7. Ions at $M/z = 17.5$ and 18.5 (total of about 2.1 pA) due to $^{35}Cl^{2-}$ and $^{37}Cl^{2-}$, and at $M/z = 9.5$ (0.25 pA) due to $^{19}F^{2-}$ are detected upon formation from CF_3Cl in the omegatron. An $^{19}F^-$ ion current of nearly 2 pA also is observed in this partial mass

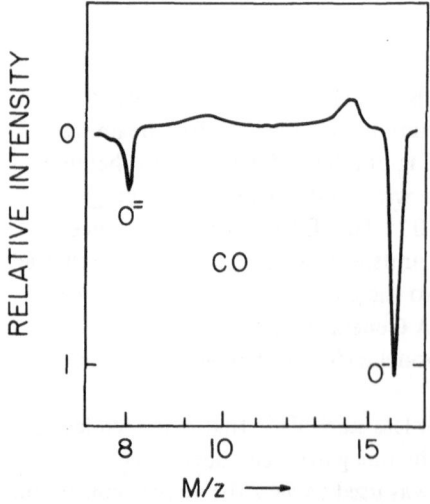

Fig. 8. Partial negative ion mass spectrum of carbon monoxide at 70 eV. (After Stuckey[280].)

spectrum. Note that in Figure 7 a somewhat higher value of $E_0 = 1.5$ V/cm was employed to increase the sensitivity. Figure 8 shows the O^- and O^{2-} ions formed in the ion-pair formation region from carbon monoxide in the omegatron. In similar manner, the omegatron experiments of Stuckey and Kiser detected the doubly-charged negative ion F^{2-} formed from CF_4, CF_2Cl_2, CF_3Br, CF_3CN and C_2F_5CN, the O^{2-} ion also formed from O_2 in air, and the Cl^{2-} ion (both isotopes) formed from CF_2Cl_2, $CFCl_3$, CCl_4 and CH_3PSCl_2. The two isotopes of the Br^{2-} ion formed from CF_3Br were detected in the approximate ratio of 1 : 1. Also, these investigators detected the doubly-charged diatomic CN^{2-} ion formed from both CF_3CN and C_2F_5CN. As will be discussed below, $SF_5{}^{2-}$, and $SF_6{}^{2-}$ ions have been observed in the negative ion mass spectrum of SF_6, but only after a re-examination of the original data.

This brief survey of the observation of doubly-charged negative ions must be accompanied also by the recognition that Z^{2-} species were not observed in all compounds examined by Stuckey and Kiser. For example, Stuckey and Kiser found only the F^- ion formed from PF_3, only F^- and O^- ions formed from POF_3, and only $POCl_2{}^-$ and Cl^- ions (and possibly a very small amount of $POCl_3{}^-$) formed from $POCl_3$. In each instance, a search for the appropriate X^{2-} or Z^{2-} ion was unrewarded. The positive ion mass spectra of these compounds were observed to contain numerous ions, and these included several doubly-charged positive ions.

It has been seen from Figs. 2 and 4, and from Eq. (4), that the flight time of the ions to the detector is of the order of 0.1 ms. Therefore, the lifetimes of all of the X^{2-} species observed must be of this same order of magnitude. Obviously the O^{2-} ion observed from CO cannot be that state with a lifetime of 5×10^{-16} s calculated by Herrick and Stillinger[136]. Since the early work of Stuckey and Kiser, other investigators also have observed doubly-charged negative ions. Their observations and experiments will be described and discussed later; however, their results indicate that the lifetimes of the X^{2-} species are $\geqslant 10^{-5}$ s.

It is important to attempt to understand the mode of formation of the doubly-charged negative ions. Singly-charged negative ions are formed typically in a relatively low energy region by resonant electron capture

$$AZ + e \longrightarrow AZ^- \tag{5}$$

and/or dissociative resonant capture

$$AZ + e \longrightarrow A^- + Z \tag{6}$$

Negative ions also are formed by electron bombardment in an ion-pair formation process at rather higher electron energies[167]

$$AZ + e \longrightarrow A^+ + Z^- + e \tag{7}$$

Therefore it was essential for Stuckey and Kiser to determine the ionization efficiency curves for the negative ions, both singly- and doubly-charged, that they had observed. These investigations would yield additional information about the formation of the long-lived doubly-charged negative ions.

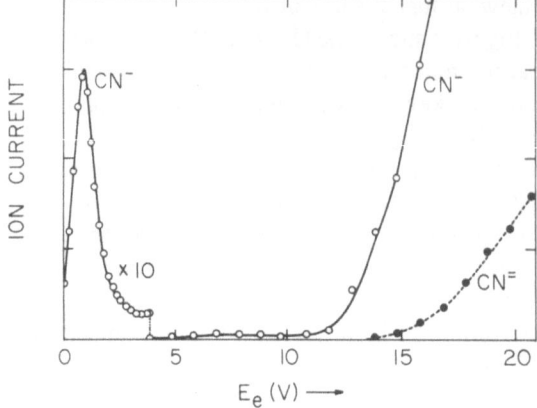

Fig. 9. Ionization efficiency curves for CN⁻ and CN2⁻ ions formed from CF₃CN. (After Stuckey[280].)

The result of such a study of the CN^- and CN^{2-} ions formed in the mass spectrum of CF_3CN is shown in Fig. 9. The CN^- formed by the dissociative resonant capture process is the peak centered at just over 1 eV. The CN^- ion-pair formation process is observed to begin at an electron energy of about 11 eV. (The energy scale is uncorrected in Fig. 9.) By investigating the intensity of the M/z = 13 ion as a function of the electron energy, E_e, the CN^{2-} ion formation was found to occur in the ion-pair formation region where CN^- ions also were observed. However, to the limit of detection sensitivity for the M/z = 13 ion, no CN^{2-} formation was observed in the resonant capture region at about 1 eV.

Investigation of the effects of pressure also should yield data useful in characterizing the mode of formation of the doubly-charged negative ions. No detailed study of the pressure dependence of CN^{2-} and CN^- ion intensities was made by Stuckey[280], although he noted the negative ion intensities were attenuated at higher pressure. However, this author notes that, from the data in Fig. 9, the intensity ratio of CN^{2-}/CN^- in the first 5 eV or so above threshold in the ion-pair formation region is linearly dependent on the CN^- ion intensity and independent of the electron energy. This observation suggests that the intensity of the CN^{2-} ion is dependent upon the square of the sample pressure.

Consider the possible ion-pair formation process

$$CF_3CN + e \longrightarrow CF_3^+ + CN^- + e \tag{8}$$

for which the intensity of the CN^- ion is given by

$$[CN^-] = k[CF_3CN][e] \tag{9}$$

Also it is important to consider the three-body electron attachment

$$CN^- + e + CF_3CN \longrightarrow CN^{2-} + CF_3CN \tag{10}$$

for which the intensity of the CN^{2-} ion is given by

$$[CN^{2-}] = k'[CN^-][CF_3CN][e] \tag{11}$$

The ratio of the CN^{2-} ion intensity to that of the CN^- ion is then

$$[CN^{2-}]/[CN^-] = (k'/k)[CN^-] \tag{12}$$

in agreement with the experimental observations. The CN^{2-} ion intensity is given by

$$[CN^{2-}] = k'k[CF_3CN]^2[e]^2 \tag{13}$$

which indicates that the CN^{2-} ion intensity is to be dependent upon the square of the CF_3CN sample pressure and upon the square of the electron beam current.

A different result was found from the ionization efficiency curves for CF_4. The dissociative resonance capture formation of F^- from CF_4 was found to have a maximum intensity at 7.1 eV (uncorrected). Dibeler, Reese and Mohler[89] indicate this process occurs at 4.5 ± 0.3 eV. The ratio of the ion currents of the F^{2-} and F^- ions formed in this resonance region was found to be essentially independent of the F^- intensity [280], in contrast to the observations just discussed for CN^{2-} from CF_3CN in the ion-pair formation region.

The effect of electron beam current on the ion intensities was investigated in some detail for the negative ions produced at 70 eV from CF_3Cl[280]. Figure 10 presents the data determined for electron trap (beam) currents of 2, 3 and 4 microamperes. It is observed that the F^- ion current increases according to the first power of the trap current. Also clear from Fig. 10 is that the Cl^{2-} ion current increases more rapidly than the F^- ion current. Although not many data are available to be considered here, reasonably straight lines that pass through the origin are obtained for the $^{35}Cl^{2-}$ and for the $^{37}Cl^{2-}$ data when the square root of the ion current is

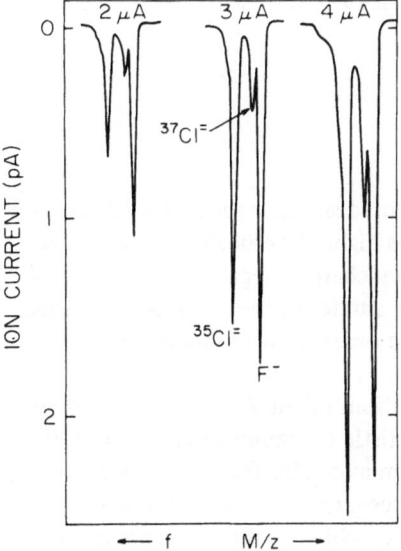

Fig. 10. Effect of the electron beam current on the intensities of F^- and Cl^{2-} ions formed CF_3Cl. (After Stuckey[280].)

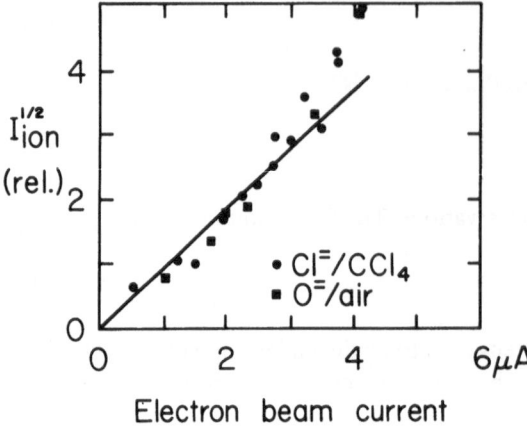

Fig. 11. Square root of O^{2-} and Cl^{2-} ion currents as a function of the electron beam current. (After Stuckey[280].)

plotted as a function of the electron beam current. The plot of the F^- ion current as a function of the electron beam intensity also yields a straight line passing through the origin. However, the plot of the square root of the F^- ion current as a function of the electron beam current does not yield a straight line, and if one does attempt to pass a "best straight line" through the data, that line does not pass through the origin. Clearly, the Cl^{2-} ion current from CF_3Cl, even as predicted for the CN^{2-} ion current from CF_3CN, exhibits a dependence on the square of the electron beam current.

This result holds equally well for other Z^{2-} ions studied. Fig. 11 summarizes a more extensive study of the square root of normalized ion intensities of both the O^{2-} ion (from O_2 in air) and Cl^{2-} (from CCl_4) as a function of the electron beam current. It is clear that, over the region of electron beam currents employed in this study, the Z^{2-} ion intensities depend upon the square of the electron beam intensity. That is, the mechanism of the formation of the Z^{2-} species must involve the attachment of two electrons. Since the Z^{2-} ion has two electronic charges, this is not a surprising result.

C. Mechanisms and Explanations

A mechanism that would have two electrons captured simultaneously by AZ to form Z^{2-} and A was dismissed by Stuckey and Kiser as highly improbable. In substantiation of that conclusion was the fact that no doubly-charged negative ions had been noted in previous negative ion mass spectrometric studies of these compounds, even with electron beam currents greater by nearly two orders of magnitude than the electron beam current used by Stuckey and Kiser.

Stuckey and Kiser[281] noted that the observations of the Z^{2-} ions were unique to the omegatron (at that time), with its characteristic of trapping ions and electrons in a relatively cool plasma about the electron beam axis. This feature provides the opportunity for many reactions to occur in the omegatron source. Ions not in resonance with the rf field could form other species which, if they then possessed the correct m/q ratio, may become resonant and be detected. Therefore, several other mechanisms were considered that recognized the unique characteristics of the omega-tron.

A more general mechanism for the production of Z^{2-} from AZ may be formu-lated; the singly-charged negative ion production may occur by a dissociative re-sonant capture reaction

$$AZ + e \longrightarrow Z^- + A \qquad (14)$$

and/or an ion-pair formation process

$$AZ + e \longrightarrow Z^- + A^+ + e \qquad (15)$$

and this may be followed by electron attachment of the Z^- ion to form the Z^{2-} ion:

$$Z^- + e + AZ \longrightarrow Z^{2-} + AZ \qquad (16)$$

Stuckey and Kiser considered this and related mechanisms and recognized that Z^- in its ground state would not bind an electron when Z is F, Cl, Br, O or CN. Therefore, they concluded that it was very unlikely that Z^{2-} formation occurs by a simple electron-capture process with the ground state of Z^-, *i.e.*,

$$Z^- + e \nrightarrow Z^{2-} \qquad (17)$$

even in the presence of a third body. Necessarily, then, an excited state of Z^- must be involved, and therefore it is convenient to consider a step involved in the for-mation of some Z^{-*} species, such as

$$Z^- + Q \longrightarrow Z^{-*} + Q \qquad (18)$$

where Q is some third body, so that the electron attachment process

$$Z^{-*} + e + Q \longrightarrow Z^{2-} + Q \qquad (19)$$

may occur and the Z^{2-} ion might be bound with respect to Z^{-*}. In some ways this is similar to He^- formation from He^{-*} discussed earlier.

There is a difficulty involved in employing Reaction (18) for a neutral Q species. Although the excitation energy in forming Z^{-*} from Z^- is unspecified, we shall see later that it is of the order of the electron affinity. This energy may well be > 2 eV, and therefore Reaction (18) is less likely than a process whereby an excited Z^{-*} species might be formed directly by reactions equivalent to those given by Reactions (14) and (15).

Stuckey[280] also considered two other mechanisms. One involved the formation of a compound negative ion by reaction with a third body

$$Z^- + Q \longrightarrow [QZ^-]^* \qquad (20)$$

and the subsequent dissociative electron capture by the compound negative ion to yield the Z^{2-} species:

$$[QZ^-]^* + e \longrightarrow Z^{2-} + Q \tag{21}$$

Using arguments concerning the electron beam energies, Stuckey found that this mechanism could be employed to explain the experimental observations.

The other mechanism considered involved the reaction of two negative ions with charge transfer occuring to form the doubly-charged negative ions, *e.g.*,

$$CN^- + F^- \longrightarrow CN^{2-} + F \tag{22}$$

and/or

$$CN^- + F^- \longrightarrow F^{2-} + CN \tag{23}$$

for the experiments that were conducted with CF_3CN, or

$$Cl^- + Cl^- \longrightarrow Cl^{2-} + Cl \tag{24}$$

in the studies of CCl_4.

Not discussed was the possibility of positive ion interactions with negative ions. Of the three mechanisms considered, Stuckey and Kiser discarded that one involving simple electron capture by Z^- in its ground state, as noted above. But they believed that the other two possibilities described should be retained for further consideration and subjected to additional experimental tests.

Another experimental finding by Stuckey also must be explained by any satisfactory mechanism for the formation of the doubly-charged negative ions. The low energy ionization efficiency curves obtained[280] for SF_6^-, Cl^- and Cl^{2-} from a mixture of SF_6 and CCl_4 are shown in Fig. 12. The same Cl^- and Cl^{2-} ion intensity

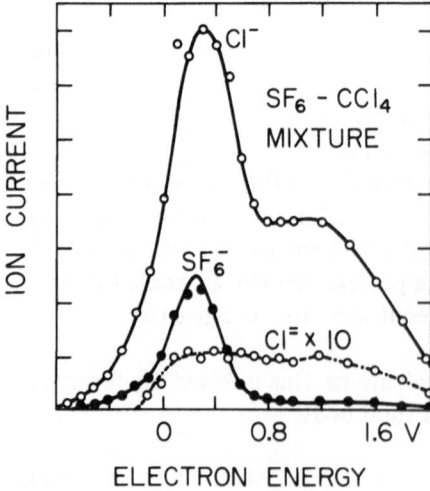

Fig. 12. Low energy ionization efficiency curves for SF_6^-, Cl^- and Cl^{2-} from a mixture of SF_6 and CCl_4 (After Stuckey[280].)

dependence on electron energy was observed when pure CCl_4 was used. Correcting the energy scale by assigning the peak maximum for SF_6^- equal to 0.0 eV[53, 73], the peak maxima for the Cl^- ions occur at 0.1 eV and 1.0 eV. The full width at half-maximum for SF_6^- is 0.45 eV, and principally reflects the energy distribution of the electron beam; that for the more intense Cl^- peak is 0.70 eV, and for the other Cl^- peak is 1.0 eV. The ratio of intensities for these two resonant capture Cl^- peaks is 2. These results may be compared with differences of < 0.6 to > 1.0 eV for the separation of the two Cl^- peaks, and ratios of $< 1.$ to *ca.* 10. reported in the literature[12, 53–55, 57, 110, 141, 193, 246]. The intensity ratio has been shown to be temperature dependent[110, 141]. The Cl^{2-} ion data mimic the Cl^- ion data, although exhibiting an intensity of only about 2 % of that of the Cl^- ions in this energy region.

Within the experimental errors in the measurements, the intensity ratio of Cl^{2-}/Cl^- from the SF_6-CCl_4 mixture was found to be constant and independent of the Cl^- intensity over the range shown in Fig. 12. This result is in agreement with that found for CF_4 in the dissociative resonant capture region, as mentioned above, and differs from the observations of CN^{2-} formed in the ion-pair formation region. Further, whereas the Cl^{2-}/Cl^- intensity ratio is constant within each of the capture peaks, the Cl^{2-}/Cl^- ratio for the peak at 1.0 eV is twice as great as the Cl^{2-}/Cl^- ratio for the peak at 0.1 eV. Stuckey[280] offered no explanation of this feature and it was not mentioned by Stuckey and Kiser[281] in their brief report of the discovery of doubly-charged negative ions.

As indicated earlier, Berry's treatment of the ion trajectories in the omegatron indicates that harmonic effects do not exist[29]. Many investigators have used and studied the omegatron mass spectrometer as a leak detector[27, 276], partial pressure analyzer[7, 60, 170, 248, 317], and as a residual gas analyzer[41, 180, 238, 247, 294, 295, 300, 304, 305]. No note has ever been made concerning the experimental observation of harmonic peaks by any of these workers.

Edwards[101] has investigated the properties of a simple omegatron. Steckel-macher and Buckingham[277] have reported on the design and evaluation of a precision omegatron system. Brodie[51] has studied ion resonance mass spectroscopy using magnetic scanning rather than using the usual fixed magnet method. Hebling and Lichtman[129] employed an omegatron to study fragmentation patterns and appearance potentials. Masica[194] has investigated the use of a simple omegatron with additional trapping voltages. Klopfer and Schmidt[171, 172] have considered the omegatron and its characteristics in detail for the quantitative analysis of gases. None of these workers report the observation of any harmonic effects.

Llewellyn[183], in reporting on the design of the first commercial ICR spectrometer, also makes no comment on harmonics. Morgan, Jernakoff and Lanneau[214] designed and developed an ion resonance mass spectrometer (an omegatron) that is essentially the same instrument Stuckey and Kiser employed. They, too, make no comment concerning any harmonic effects. Redhead[245] observed oscillations in the electron trap current of an omegatron and McNarry[205] has investigated these oscillations related to the ion plasma confined by a magnetic field. These oscillations were found to be independent of the frequency and are not to be confused with harmonics. All of these experimental findings are in accord with the theoretical treatment[29].

In summary, no evidence has ever been presented for harmonic effects in the omegatron mass spectrometer. Stuckey and Kiser explicitly state that they were unsuccessful in detecting any harmonic effects. They point to several experimental results which negate any resort to "harmonic effects" to explain the peaks the observed and attributed to X^{2-} ions. Three of these are:

(1) They utilized the oscilloscope to monitor continuously the rf frequency applied to the omegatron field plates and were unable to detect any harmonics present in the applied rf field;

(2) they made a careful effort to observe CN^{2-} in the dissociative resonant capture region of CN^- from CF_3CN and could not detect the CN^{2-} ion there, but did detect the CN^{2-} ion formed from CF_3CN in the ion-pair formation region (see Fig. 9); and

(3) they examined the mass spectrum of air in the $M/z = 6$ to 9 region with an intermediate trapping voltage that permitted observation of both positively- and negatively-charged ions. These data are shown in Fig. 13. Clearly observed are both O^{2+} and O^{2-} ions, but only the N^{2+} ion; an N^{2-} ion is not observed. If the O^{2-} ($M/z = 8$) peak were to be ascribed to a harmonic effect, surely that same effect must cause a peak to occur where N^{2-} ($M/z = 7$) is to be expected.

The conclusions drawn by Stuckey and Kiser were as follows:

(1) The ions observed at $M/z = 17.5$ and 18.5, at $M/z = 9.5$, and at $M/z = 39.5$, and 40.5 are due to Cl^{2-}, F^{2-}, and Br^{2-}, respectively. Further, the ions found at $M/z = 8.0$ in the study of the oxygenated compounds and at $M/z = 13.0$ in the investigation of CF_3CN are due to O^{2-} and CN^{2-}, respectively;

(2) the observation of the Z^{2-} ions cannot be attributed to harmonic effects;

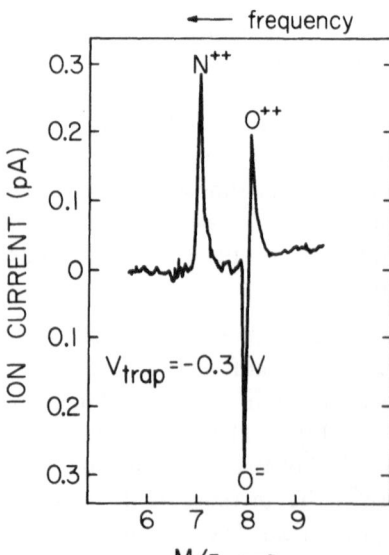

Fig. 13. N^{2+}, O^{2+} and O^{2-} ions observed when formed from air using an intermediate trapping voltage of -0.3 V. (After Stuckey[280].)

(3) the lifetimes of these doubly-charged negative ions are of the order of 0.1 ms;

(4) the formation of the doubly-charged negative ions are to be attributed to a unique characteristic of the omegatron and very likely excited states are involved in their formation and account for the long lifetimes;

(5) the Z^{2-} ion intensities depend upon the second power of the electron beam current; and

(6) although several mechanisms could be discarded as inadequate or unlikely, at least two mechanisms remained as possibilities for further consideration.

As shall be seen in the following pages, experimental evidence to support the existence of these, and other, doubly-charged negative ions has recently been reported. And it is considered quite probable that many other doubly-charged negative ions remain to be observed under the appropriate experimental conditions.

Fremlin[113] recognized the importance of the experiments of Stuckey and Kiser, but was concerned with the possibility that the observed peaks might be artifacts produced by excitation of harmonics in spite of Stuckey and Kiser's comments about the non-existence of harmonic effects.

Fremlin proposed two mechanisms in an attempt to account for the observations of Stuckey and Kiser. In essence, both of these mechanisms were attempts to invoke harmonic effects. The second mechanism proposed did not appeal even to Fremlin, and since ICR experiments[275] have detected doubly-charged negative ions also (see discussions below), it can now be eliminated from further consideration.

The first mechanism suggested by Fremlin contains several assumptions, contingencies and uncertainties. If neutral atoms are formed as Fremlin suggests, and these neutral atoms are the basis for the Z^{2-} ion observations, then "Z^{2-} ions" should be observed at all electron energies where *any* negative ions are formed, and for the resonant frequencies of all negative ions. These predictions are not in agreement with the experimental facts reported by Stuckey and Kiser[281].

Fremlin's retarding potential experiment and pulsing experiment are both worthwhile suggestions; Stuckey and Kiser had not investigated these, for they had not found any harmonic effects. Nonetheless, these experiments should be conducted to provide additional evidence that the mechanisms Fremlin considered are not operative. The pressure dependency experiments, suggested by both Fremlin and Stuckey, are needed. Baumann, Heinicke, Kaiser and Bethge[26, 34] and others have conducted qualitative pressure studies, but more detailed investigations would be desirable.

Chupka, Spence and Stevens[63], in recognizing the fundamental importance of the long-lived doubly-charged negative ions reported by Stuckey and Kiser, also have expressed the belief that the observed ion peaks may be experimental artifacts.

Although few details of their experiments are provided, Chupka, *et al.* indicated that they were unable to obtain any evidence for the existence of X^{2-} ions from I_2, HI, CF_3Cl, CH_3I and an H_2-I_2 mixture with an electron bombardment source and a double-focusing mass spectrometer of 280-cm radius. Apparently they did not recognize that the omegatron source is not a simple electron bombardment source. Thus, their experiments did not adequately duplicate the omegatron experiments (or the ICR experiments, to be discussed below) and cannot be interpreted as indicating that Z^{2-} ions do not exist.

Since the discovery of doubly-charged negative ions by Stuckey and Kiser, a number of other investigators also have reported observations of Z^{2-} species. It is appropriate to consider their experiments and observations. Also, it is instructive to compare the species observed and the proposed methods for formation with those reported by Stuckey and Kiser, for this gives additional insight into the mechanism(s) of formation of the doubly-charged negative ions.

III Observation of Doubly-Charged Negative Ions from a Penning Source Using a Conventional Mass Spectrometer

In 1971, Baumann, Heinicke, Kaiser and Bethge[26] reported that they had detected doubly-charged negative ions during an investigation of the properties of negatively charged atomic and molecular ions produced from a Penning source. The mass spectrometer employed was a 60° magnetic sector with an electric field positioned after the magnetic sector to deflect the ion beam to one of two Faraday cup detectors. This first confirmation of the existence of doubly-charged negative ions merits a detailed review, particularly with respect to the ion source employed, the experimental approaches and the results obtained.

A. Penning Ion Source

F. M. Penning[234, 235] designed an ionization gauge to operate at and measure pressures down to about 10^{-6} Torr. In 1949, Penning and Nienhuis[237] reported the design of a substantially improved "Phillips Ionization Gauge", also commonly referred to as a "Penning Gauge"[240]. These ionization gauges, particularly the improved design that used a cylindrical anode rather than the original loop anode, were the basis for the cold-cathode Penning ion source, first described by Penning and Moubis[236]. In the Penning and Nienhuis gauge the discharge space is enveloped by

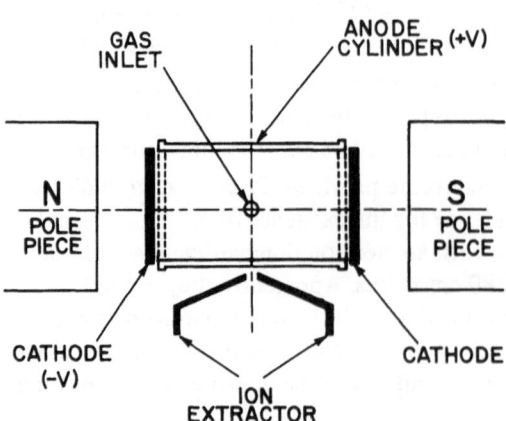

Fig. 14. Schematic diagram of a penning ion source with radial ion extraction

the electrodes and is no longer affected by the metallic layer deposited on the glass walls by cathode sputtering. A schematic diagram of this type of ion source, with radial ion extraction, is shown in Fig. 14. It is seen that there is at least a superficial resemblance of this source and the omegatron design described earlier.

Ion collisions at either of the cathodes cause secondary electron emission. These electrons are accelerated by a sizeable potential drop toward the center of the anode but are constrained by the magnetic field to pass through the anode cylinder rather than travelling directly to the anode. The paths of electrons not initially normal to the anode axis are necessarily helical. As the electrons pass through the anode they are slowed down and reflected before the other plane cathode. Thus, the electrons are caused to execute oscillatory helical trajectories back and forth through the cylindrical anode. The trajectories of the electrons have been considered by Hull[155]. As a result of the use of a magnetic field, the electron paths through the gas in the source are much longer (more than 100 times the distance between the two cathodes[240]) than if no magnetic field were present, and the probability that they will cause ionization and undergo electron attachment is very high.

The discharge can be maintained with a voltage of about 2 kV in a magnetic field of some 400–1000 G at pressures $\ll 10^{-3}$ Torr. The large surface area of the cathodes provides a substantial current. Although sputtering limits the life of the Penning source, careful choices of electrode materials and discharge support gases can permit these sources to have useful lifetimes of the order of several hundred hours[130]. Lorrain[185], Keller[164], Böhm and Günther[42], and Bethge, Heinicke and Baumann[35, 130, 131] have studied and designed improved versions of the Penning ion source for various applications. Lorrain[185] has pointed out that there is a similarity of this type of ion source with that devised earlier by Maxwell[199]. Also, as Dushman and Lafferty[98] point out, the inverted magnetron due to Redhead[245] has efficient electron trapping in the discharge region and is related to the Penning ion source.

B. Experimental Arrangement and Techniques

The specific Penning ion sources employed in the study of the doubly-charged negative ions have been described[25, 35, 130, 131]. In a typical experiment involving a study of Te^{2-} [34], the operational parameters of the Penning source (developed further from the basic design of Keller[164]) were as follows: tellurium vapor was fed into the discharge maintained at about 10^{-3} Torr by hydrogen gas. An arc voltage of 1.8 kV, an arc current of 50 mA, and a magnetic field of about 400 G were employed. Ion energy spreads of such an ion source are typically of the order of 10 eV or more.

The ions produced in the dense plasma of the Penning ion source are extracted and, after acceleration through a 20 kV potential drop, are focused by passage through an arrangement of einzel lenses and deflected by the magnetic field, as shown in Fig. 15. The pressure in the accelerating and analyzing regions was less than about 5 x 10^{-6} Torr[26, 34]. With a resolution of about 150, the electrically undeflected ion beam was detected by Faraday cup 1. When electrical deflection was

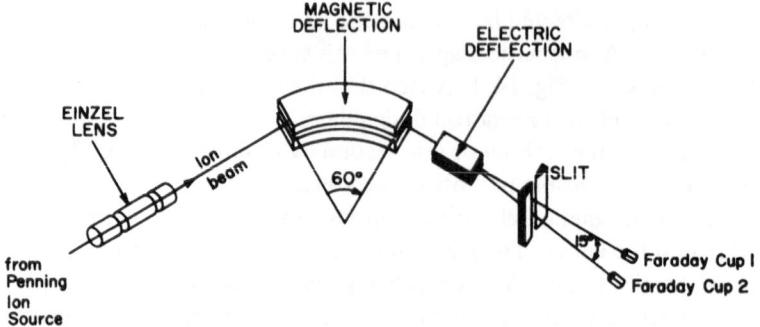

Fig. 15. Experimental arrangement used by Baumann, Heinicke, Kaiser and Bethge in the observations of doubly-charged negative ions[26)]

employed the ion beam was detected by Faraday cup 2, inclined about 15 ° to the ion beam axis.

A magnetic field is used to provide momentum/charge-analysis of the ion beam according to the expression

$$mv = qrB \tag{25}$$

where m is the mass and q is the charge of the ion moving in a circular path of radius r in the magnetic field, B. The angle of deflection of an ion by the magnetic field is given by

$$\theta = qsB(2mT)^{-0.5} \tag{26}$$

where s is the length of the arc of the circular path of the ion in the magnetic field and T is the kinetic energy which the ion acquires in the accelerating system by passing through a potential drop of V,

$$T = mv^2/2 = qV \tag{27}$$

Alternatively, Eq. (26) may be written as

$$\theta = sB(q/2mV)^{0.5} \tag{28}$$

For a fixed magnetic deflection angle, the peaks observed in the mass spectrum may be identified from values of B, r and V with

$$m/q = B^2 r^2/2V \tag{29}$$

Figure 16 presents the mass spectrum obtained with iodine in the Penning ion source and no electric deflection employed. The peak observed at M/z = 63.5 may be due to I^{2-} and/or to the daughter ion of a metastable transition occurring prior to the magnetic sector,

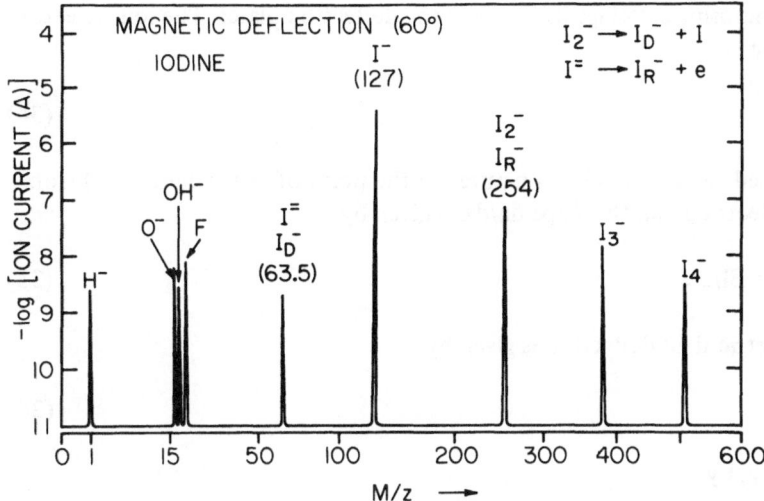

Fig. 16. Mass spectrum of iodine obtained by magnetic analysis. (After Baumann, Heinicke, Kaiser and Bethge[26].)

$$I_2^- \longrightarrow I(D)^- + I \tag{30}$$

designated as $I(D)^-$ by Baumann, et al.[26].

The peak detected at M/z = 127 principally is due to I^-, but may contain a small contribution due to the daughter ion of the metastable transition

$$I_4^- \longrightarrow I_2(D)^- + I_2 \tag{31}$$

The peak at M/z = 254 may be composed both of I_2^- and $I(R)^-$, the heavy product of the electron detachment of an I^{2-} ion prior to the magnetic sector,

$$I^{2-} \longrightarrow I(R)^- + e \tag{32}$$

The magnetic sector will not separate the possible components of the M/z = 63.5 peak or of the M/z = 254 peak. The Heidelberg group of Baumann, Heinicke, Kaiser and Bethge[26] used the electric deflection plates for a kinetic energy analysis in a manner similar to that employed by several other investigators for studies of the metastable transitions of ions[15, 37, 38, 78, 121, 168, 249, 260] with double-focusing mass spectrometers. Here, however, Baumann, et al.[26] electrically deflected the ion beam perpendicular to the plane employed in the momentum analysis rather than in the same plane.

With parallel plates of length l in the x-direction (the direction of the ion beam), separated by a distance d in the z-direction, a uniform electric field of

$$E = U/d \tag{33}$$

is formed by imposing a voltage of U across the deflection plates. The ion trajectory then is given by

$$z = (E/4V)x^2 \tag{34}$$

and is recognized to be parabolic in nature. At the point of exit (where $x = 1$) from this uniform electric field, the slope dz/dx is given by

$$dz/dx = \tan \phi = El/2V \tag{35}$$

so that the electric field deflection is given by

$$\tan \phi = Ul/2dV \tag{36}$$

or, alternatively, by

$$\tan \phi = qUl/2dT = qUl/dmv^2 \tag{37}$$

from which it is seen that the analysis is independent of the mass of the charged particles but is dependent upon the kinetic energy of the ions.

The meaning of these expressions is made clearer by the following example. The kinetic energy of an I_2^- ion accelerated through V is T. If the I_2^- ion decomposed according to Reaction (30), i. e., after its acceleration and passage through the einzel lens, but before the momentum/charge-analysis, the I^- decomposition product, designated as $I(D)^-$, has the same velocity as the undecomposed I_2^- ions but only one-half of the mass of I_2^- and so will be momentum/charge-analyzed along with any I^{2-} ions. Since the kinetic energy of $I(D)^-$ is T/2 whereas that of I^{2-} is T, the electric deflection will separate these two components of the $M/z = 63.5$ ion beam. If a (critical) voltage of U_c is required to deflect the I^{2-} ions (of kinetic energy T) through the angle ϕ to Faraday cup 2, then a voltage of $U_c/2$ will deflect the $I(D)^-$ ions (of kinetic energy T/2) through this same angle. Thus, by varying the voltage U across the deflection plates, the momentum/charge-analyzed beam may also be energy analyzed.

C. Results and Discussion

The results[26] of an electric deflection experiment for the momentum-charge-analyzed negative ions of $M/z = 63.5$ formed from iodine added to the Penning source are illustrated in Fig. 17. The example chosen above obtains in Fig. 17, and it is observed that the $M/z = 63.5$ peak observed in the momentum/charge-analysis (see Fig. 16) is composed of two major peaks, even as the above example indicates, and a rather smaller third peak. The peak found at $U_c/2$ is $I(D)^-$ and that at U_c is due to the doubly-charged negative iodine ion.

The much smaller third peak observed at $2U_c$ in Fig. 17 is due to an apparent increase in kinetic energy to 2T. This peak at $2U_c$ is attributed to an electron detach-

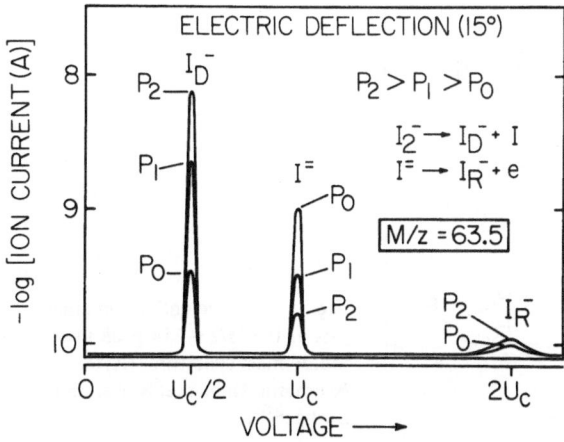

Fig. 17. Electric deflection analysis of the M/z = 63.5 peak in the iodine mass spectrum. (After Baumann, Heinicke, Kaiser and Bethge[26]).

ment reaction of the I^{2-} ion, according to Reaction (32). An I^{2-} ion, of kinetic energy T and velocity v, that undergoes electron detachment after the einzel lens and prior to the magnetic sector, will form the I^- ion, designated as $I(R)^-$. The kinetic energy of $I(R)^-$ remains very nearly T. The momentum of the $I(R)^-$ ion is the same as that of the I^{2-} ion. But since the charge of the $I(R)^-$ ion no longer is equal to 2e, it will not be transmitted by the magnetic sector at an angle of θ that is the same as that for I^{2-} and $I(D)^-$.

However, these reactions are not limited to the first field-free region that follows the acceleration and einzel lens region and precedes the magnetic field. Reactions may occur in the magnetic (and also in the electric) sector, but these generally give rise to very low intensity ionic contributions over a wide and near-continuous range of apparent M/z and therefore are very difficult to detect and analyze[111]. We shall neglect further consideration of these here. However, those reactions which may occur in the second field-free region, between the magnetic and electric sectors, must be considered.

For the I^{2-} ion that is accelerated and momentum/charge-analyzed with no further reactions, a value of U_c will be required to deflect it through ϕ to Faraday cup 2, as given by Eq. (37). If, however, I^{2-} detaches an electron in the second field-free region, the charge on the resultant species is decreased. Thus, although m and v are the same for I^{2-} and $I(R)^-$, the values of q for these two species are not the same. Since U is inversely proportional to q for a constant ϕ, the decrease from q = 2e to q = e associated with the process given by Reaction (32) means that $I(R)^-$ formed in the second field-free region will be detected by Faraday cup 2 at $2U_c$.

Similar arguments apply to the results (see Fig. 18) of the electric deflection analysis of the M/z = 254 ion beam. Both I_2^- and the I^- formed as a result of the repulsive electron detachment from I^{2-} in the first field-free region (that is, $I(R)^-$) will be transmitted by the magnetic sector. Kinetic energy analysis of this ion beam requires $2U_c$ to bring the $I(R)^-$ species to Faraday cup 2 and U_c to detect the I_2^- ion. Any I^- formed by dissociation of $I2^-$ in the second field-free region, $I(D)^-$, will be detected at Faraday cup 2 by application of $U_c/2$ to the electric deflection plates.

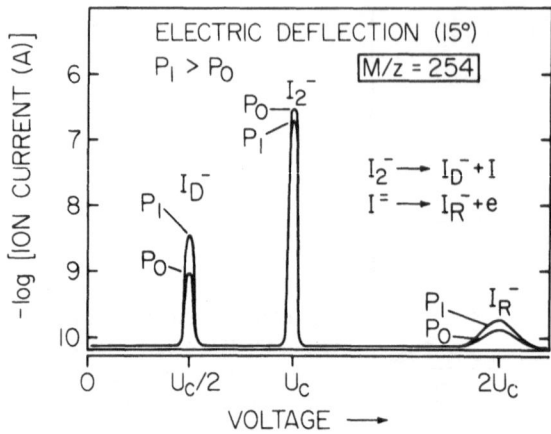

Fig. 18. Electric deflection analysis of the M/z = 254 peak in the iodide mass spectrum. (After Baumann, Heinicke, Kaiser and Bethge[26].)

Specific results for tellurium, similar to those shown in Figs. 16 through 18 for iodine, have been reported by Bethge[34]. In this manner, Baumann, Heinicke, Kaiser and Bethge[26, 33] have detected O^{2-}, F^{2-}, Cl^{2-}, Br^{2-}, I^{2-}, Te^{2-}, and Bi^{2-} formed in their Penning ion sources. Bethge[34] has noted that P^{2-}, As^{2-} and Sb^{2-} have been found to exist also. Thus, this work appears to corroborate the existence of the doubly-charged monatomic species first reported by Stuckey and Kiser[280, 281]. It also provides additional information about the doubly-charged negative monatomic ions, namely, (1) they do undergo the expected electron detachment process, and (2) their intensities are affected by pressure.

Stuckey and Kiser did not report a pressure study, so that the qualitative results of Baumann, Heinicke, Kaiser and Bethge[26] and Bethge[34] are quite important. From Figs. 17 and 18 it can be observed that as the pressure in the beam handling system is increased, the intensity of $I(D)^-$ increases while the intensities of $I(R)^-$ and I^{2-} decrease significantly. The same is true of the corresponding Te species[34]. This indicates that I^{2-} and Te^{2-} (and presumably all Z^{2-}) ions undergo collision-induced electron detachment even as the I_2^- and Te_2^- undergo collision-induced dissociation.

It is interesting to recognize that the beam currents of the Z^{2-} species reported by Baumann, Heinicke, Kaiser and Bethge[26] and Bethge[34] range from about 0.2 nA to greater than 2 nA, whereas those detected by Stuckey and Kiser were several hundred times less intense, e. q., about 2 pA of the Cl^{2-} ion were reported formed from CF_3Cl. Baumann, et al.[26] note that the Z^{2-} beam currents increase with increasing mass of the elements involved and Bethge[34] suggests as an explanation that "a large shell system can easily be deformed to provide binding through the induced dipole moment" for the Z^{2-} species.

Bethge[33] indicated in 1972 that he did not know either the electronic configurations of the Z^{2-} species or the lifetimes of these doubly-charged negative ions. However, he did suggest that the configurations resembled those of the alkali metals. In a 1973 publication, Bethge[34] noted that since the ion flight times were $> 10^{-7}$ s, the Z^{2-} lifetimes were $> 10^{-7}$ s. In this report, Bethge noted the study of Marmet, Bolduc and Quéméner[192] and suggests that two electrons must be excited simultaneously to form the Z^{2-} species.

Of course, successful duplication of the experiments of the Heidelberg group would be most welcome. Chupka, Spence and Stevens[63] indicated several years ago that they were attempting to use a Penning ion source and a double-focusing mass spectrometer in an effort to verify the observations made by Baumann, Heinicke, Kaiser and Bethge[26]; however, this author is unaware of any published results by Chupka, et al., of the findings of their work. Nor is the author aware of a successful application of doubly-charged negative ions in a tandem accelerator by Bethge and co-workers.

In a recent review of the binding energies of atomic negative ions, Hotop and Lineberger[153] report that D. Feldmann has been unsuccessful in using the same ion source that Baumann, Heinicke, Kaiser and Bethge employed to produce a beam of doubly-charged negative ions that could be used for photodetachment studies. Since no experimental details and results have been published, it is impossible as of the present to attach any significance to this report.

IV Other Observations of Doubly-Charged Negative Ions

Several other observations of a variety of doubly-charged negative ions have been made by different investigators using different experimental approaches. The experimental methods employed do, however, have some common bases. The experiments described below involve different ion sources, but these can be classified into two main categories: conventional Nier-type ion sources and relatively cool plasma sources. The experiments of Ahnell and Koski[4, 5], Dougherty[94] and Bowie and Stapleton[47] involve the Nier-type ion sources. The experiments of Schnitzer and Anbar[10, 255, 256] and Stapleton and Bowie[275], like those of Baumann, Heinicke, Kaiser and Bethge[26] and Stuckey and Kiser[280, 281] involve plasma-type ion sources, although of somewhat different types, temperatures and densities. The following discussions will outline briefly each of the experiments and relate the interpretations wherever possible.

A. Observation of F^{2-} with a Quadrupole Mass Spectrometer

Ahnell and Koski[4, 5] have reported the observation of F^{2-} ions from CF_3Cl, one of the molecules employed by Stuckey and Kiser[280, 281] to study the F^{2-} ion. The mass spectrometric instrumentation used by Ahnell and Koski[4, 5] differs from those described above and therefore their observations and results merit careful consideration.

Ahnell and Koski report using a monoenergetic beam of electrons (40 mV FWHM) to bombard the target molecules in the ion source. The mass spectrometric system was one designed to study ion-pair processes by detecting both the M/z-analyzed positive and negative ions in coincidence. Svec and Flesch[282] have previously reported on such a system using sector magnetic field instruments. The mass spectrometers used by Ahnell and Koski were quadrupole instruments[79, 228, 229] that could be employed individually or in coincidence.

The sample pressure, electron beam current, auxiliary magnetic field, and ion residence time in the ion source are not stated and cannot be inferred from the data given. As will be seen below, the details of the ion source are quite important in the production of doubly-charged negative ions.

A figure is presented by Ahnell and Koski[4] that clearly indicates the formation of F^{2-} from CF_3Cl at an electron energy of 100 eV. The M/z of this ion is 9.5 and its relative abundance is approximately 8 % (F^- is the base peak in their negative ion mass spectrum of CF_3Cl), although the intensity of the F^{2-} ion decreased as the electron energy was decreased. Their value of 8 % for the F^{2-} ion compares favorably with the relative abundances reported by Stuckey and Kiser. Also clearly evident in Ahnell and Koski's mass spectrum is the Cl^- ion at M/z = 35 (ca. 27 %) and M/z = 37 ca. 9 %). However, they failed to detect the Cl^{2-} ion at M/z = 175 and 18.5, although Stuckey and Kiser easily detected Cl^{2-} (with relative abundances of about 75 % for M/z = 17.5 and 25 % for M/z = 18.5 under the stated conditions) in their studies with CF_3Cl.

In addition to F^-, Cl^- and F^{2-}, the negative ions mass spectrum given by Ahnell and Koski[4] indicates a peak at M/z = 27 and hints of the presence of CF^- and ClF^- ions. Possibly other species also are present, but the low resolution of the mass spectrum shown precludes any attempt at definitive assignment by this author. The presence of the M/z = 27 ion (possibly HCN^- ?) with a relative abundance of about 2–3 % in their spectrum is not commented on by Ahnell and Koski.

Ahnell and Koski note that they did not find F^{2-} when CF_4 was used as the target molecule in experiments similar to those they conducted with CF_3Cl, whereas Stuckey and Kiser did observe F^{2-} from CF_4 in the omegatron experiments. The total cross section for ionization for CF_3Cl is likely to be less than a factor of two greater than that for CF_4. Craggs and Massey[69] have reported data due to Marriott and Craggs which indicate that the relative abundance of the F^- ion in the mass spectra of these two molecules to be essentially the same; similar results have been found by Hobrock[145]. Dibeler, Reese and Mohler[89] found the relative abundance of F^- from CF_4 to be about four times the relative abundance of F^- from CF_3Cl. Dibeler and Reese[88] measured the relative abundance of F^- from CH_3F to be about six times the relative abundance of Cl^- from CH_3Cl. Thus, the F^{2-} from CF_4 should be observed, as also should Cl^{2-} from CF_3Cl. The above investigators, as well as Chait, Askew and Matthews[59], indicate that F^- is the base peak in the negative ion mass spectrum of CF_4. It is noted that Svec and Flesch[282] did not report observing F^{2-} in their study of the positive and negative ions formed from chromyl fluoride.

One possible explanation for this non-observation of F^{2-} from CF_4 by Ahnell and Koski, bearing the above information in mind, is that the production of F^{2-} occurs via a different route in Ahnell and Koski's experiments than in the omegatron experiments by Stuckey and Kiser or in the studies by Baumann, Heinicke, Kaiser and Bethge[26, 33, 34] using a Penning source.

It is concluded by Ahnell and Koski that their "results indicate that the F^{2-} ion must be formed in an excited state by some indirect process in which a singly-charged negative ion reacts either with the parent molecule or some fragment of it". No other information is given concerning this conclusion.

Chupka, Spence and Stevens[63] have made an effort to verify the observation of doubly-charged negative ions produced by electron bombardment. Few details of their experiments are given. They state that they attempted to duplicate the conditions of the electron bombardment experiments, and that they failed to find any evidence of X^{2-} ions produced from I_2, HI, CF_3Cl, CH_3I and mixtures of H_2 and I_2. They conclude that they have established a lower (upper ?) limit for the intensity ratio of X^{2-}/X^- of $< 10^{-11}$. If their experiments did duplicate those of Ahnell and Koski, then the observed $M/z = 9.5$ peak still requires an explanation, although the lack of observation of Cl^{2-} from CF_3Cl and of F^{2-} from CF_4 would be in agreement with Chupka, Spence and Stevens. Further, a separate mechanism for X^{2-} formation would be obviated. Additional experiments are required to clarify and resolve these differences.

B. Observations of Doubly-Charged Negative Ions with ICR Instrumentation

Stapleton and Bowie[275] recently conducted the first study of doubly-charged negative ions by means of ion cyclotron resonance (ICR) spectroscopy. The basic principles and techniques of ion cyclotron resonance spectroscopy are quite similar to those of the omegatron, as noted above, although there are also two important differences, i. e., direct measurement of ion current versus power absorption from the exciting field at resonance, and the separation of the ion production and the ion detection regions. Because of the great similarity, and also because of the specific differences just cited, an ICR search for doubly-charged negative ions is of particular importance to the omegatron studies described previously.

Stapleton and Bowie[275] found doubly-charged negative ions in their ICR investigation after an earlier search with a conventional magnetic sector mass spectrometer by Bowie and Blumenthal[45] that was unproductive and then one by Bowie and Stapleton[47] that was reported to be rewarding. In the following paragraphs the approaches, observations and conclusions reached by Stapleton and Bowie from their ICR study will be examined.

The instrumentation employed by Stapleton and Bowie was an ion cyclotron resonance mass spectrometer that had been modified to permit computer control of all drift voltages and to allow direct reading of the ion transit time (typically 10^{-4} to 10^{-3} seconds) similar to that described by McMahon and Beauchamp[204]. An emission current of about 0.2 microampere and a nominal 70 eV electron beam produced ion currents of 10^{-12} to 10^{-11} A at source pressures of approximately 10^{-6} Torr. The mass spectra were measured by magnetic field modulation.

Double resonance and ejection spectra were measured as described by Goode, O'Malley, Ferrer-Correia and Jennings[119, 120]. These conditions were found to be quite satisfactory and permitted Stapleton and Bowie[275] to measure with greater sensitivity the doubly-charged negative ions previously observed[47] with their conventional mass spectrometer (these experiments and results are discussed below) and also to detect other doubly-charged negative ions they had not observed previously.

Specific experimental information is given by Stapleton and Bowie[275] concerning the formation of Cl_2^- from CCl_4; Stuckey and Kiser observed this ion from the same compound in their original investigations. For an ion transit time of 0.15 ms at an ion current of 10^{-11} A and using a cyclotron frequency of 153.7 kHz, the intensity of the Cl_2^- ion was found to be about 9×10^{-3} that of the Cl^- ion, dependent upon the electron energy. Also, the power absorption of the Cl_2^- ion was noted to be dependent upon the CCl_4 sample pressure. The Cl^- ion is present at 10^{-7} Torr, whereas the sample pressure had to be increased to 5×10^{-7} Torr before the Cl_2^- ion was first observed. A maximum signal corresponding to Cl_2^- was obtained at 5×10^{-6} Torr; at this pressure, the Cl_2^-/Cl^- ratio was 10^2 greater than that observed at 5×10^{-5} Torr in their conventional mass spectrometer with a trap current of nearly 10^2 microamperes.

Stuckey and Kiser found in the omegatron studies that Cl_2^- ions were produced from CCl_4 in both a low energy, resonant capture region (from about 0 to a little more than 2 eV; see Fig. 12) and in a higher energy, Cl^- ion-pair formation region (above about 10 eV). Very similar results were found by Stapleton and Bowie[275]; a maximum was found in the region of 1 to 3 eV (they found the Cl_2^-/Cl^- intensity ratio to be 10^{-3} at 2 eV whereas Stuckey and Kiser obtained a ratio of about 3×10^{-2} at 1 eV) and another maximum in the region of 35–40 eV which they attributed to dissociative secondary electron capture. Stapleton and Bowie contrast these results with those of Ahnell and Koski[4] who found, as has been discussed above, a maximum F^{2-} (not Cl^{2-} as stated by Stapleton and Bowie) ion formation at 70 eV.

A further group of results obtained by Stapleton and Bowie should be noted. Even as Baumann, Heinicke, Kaiser and Bethge[26] observed, Stapleton and Bowie found that the addition of collision gases decreased the detected intensity of the X^{2-} ion. This was so for the Cl_2^- ion formed from CCl_4, and was analogous to the behavior they observed[275] for the doubly-charged negative ions from p-nitrobenzoic acid in their conventional mass spectrometer. Stapleton and Bowie conclude that this observation discounts the operation of a collision mechanism involving a neutral species in the formation of the doubly-charged negative ions. Further comment on this point will be made later in the discussion of the mechanism of formation of the doubly-charged negative ions. Also, their[275] studies using double resonance and cyclotron ejection techniques showed that the Cl^- and Cl_2^- ions from CCl_4 are coupled. This finding substantiates the earlier suggestions of Stuckey and Kiser with regard to the formation of the Z^{2-} ions from the Z^- ions based on the ionization efficiency curves.

Finally, in addition to the observation of Cl_2^- from CCl_4, Stapleton and Bowie also observed formation of both SF_5^{2-} and SF_6^{2-} in small abundances from SF_6. Stuckey and Kiser employed SF_6 as an energy scale calibrant in their investigations of the energetics of the singly- and doubly-charged negative ions, but did not report any SF_5^{2-} or SF_6^{2-} ion formation. In their negative ion study of pure SF_6 in the omegatron, Stuckey and Kiser noted peaks at 54 and 63 kHz that corresponded to M/z = ca. 146 and 74, respectively. Two much smaller peaks (relative abundance of about 5–10 %) at 107 and 124 kHz also were observed in the spectrum. Because of the rather poor resolution at these high M/z values, these ions corresponding to

M/z = ca. 74 and 64, respectively, were not discussed in the published works[280, 281]. However, a review by this author of Stuckey's original data establishes that the peaks at 124 and 107 kHz are attributable to SF_5^{2-} and SF_6^{2-}, respectively. These data corroborate the reported observations of SF_5^{2-} and SF_6^{2-} by Stapleton and Bowie. Stuckey's 70 eV positive ion mass spectrum of SF_6 exhibits peaks due to SF_5^+ (100.), SF_4^+ (10.), SF_3^+ (30.), SF_2^+ (10.), SF_5^{2+} (40.), SF_4^{2+} (5.) and SF^+ (10.) ions; values in parentheses are the relative abundances. In this same study, Stuckey also noted that as the pressure of SF_6 was increased the negative ion intensities decreased. It should be recognized that no SF_5^{2-} or SF_6^{2-} was reported by Herzog[138] in the negative ion mass spectrum of an SF_6-H_2O mixture using a monopole mass spectrometer; however, Herzog used a Nier-type ion source in his study; use of a quadrupole ion trap might produce different results.

It was noted earlier that the experiments reported by Chupka, Spence and Stevens[63] did not duplicate the ion source conditions of the omegatron experiments[280, 281]. Similarly, the studies of Chupka, Spence and Stevens did not duplicate the conditions of the ICR experiments of Stapleton and Bowie.

C. Reported Observations of Polyatomic Doubly-Charged Negative Ions

Based on the argument that a large separation of the two charges would substantially reduce the Coulombic repulsion and thereby increase the stability and lifetime of a doubly-charged negative ion, Dougherty[94] examined the negative ion mass spectra of several polyatomic molecules. The compounds included in his investigation were benzo[cd]pyrene-6-one (I), perinaphthenone (II), and cyclo-octatetraene (III):

I   *II*   *III*

Only in the negative ion mass spectrum of the benzo[cd]pyrene-6-one was a peak observed that could be assigned as due to a doubly-charged negative ion. Interestingly, this ion was observed at M/z = 254.5 and the parent molecule has a nominal molecular weight of 254. The peak at M/z = 254.5 was consistently observed on a shoulder of the pressure-broadened parent negative ion peak at M/z = 254 and was interpreted by Dougherty as due to an ion of M/z = 509/2, where 509 is ascribed to the ^{13}C-isotopic component of the dimer of I. Any doubly-charged negative ion of the dimer of I, of course, would have M/z = 254 and be indistinguishable from the singly-charged negative ion of the monomer of I.

Let us take the isotopic abundances of ^{12}C (0.98888) and ^{13}C (0.01112) and compare the observed and expected relative intensities of the results for which data are available (the data given do not warrant inclusion of the isotopic species of hydrogen and oxygen.) From the data presented in the figure given by Dougherty[94], it is clear that the relative intensities of the ions at M/z = 254 and 255 in the positive ion spectrum correspond closely to those expected for the species $^{12}C_{19}H_{10}O^+$ and $^{13}C^{12}C_{18}H_{10}O^+$.

The relative intensities of the peaks at M/z = 254 and 255 in the negative ion spectrum do not correlate well with those expected for $^{12}C_{19}H_{10}O^-$ and $^{13}C^{12}C_{18}H_{10}O^-$; however, they do correlate rather well with $^{12}C_{38}H_{20}O_2^-$ and $^{13}C_2{}^{12}C_{36}H_{20}O_2^-$ if it is assumed that peak height is a direct measure of the ion intensity (i. e., there are no cross-contributions of the intensities of peaks, or that the resolution exceeds ca. 500.) With this same assumption the intensity of the M/z = 254.5 ion agrees very well with that to be expected for $^{13}C^{12}C_{37}H_{20}O_2^-$. On the basis of assignment of the M/z = 254.5 peak as due to $^{13}C^{12}C_{37}H_{20}O_2{}^{2-}$, essentially all of the intensities of the M/z = 254, 254.5 and 255 peaks are due to $C_{38}H_{20}O_2$, the dimer of I, and there is very little or no intensity in the M/z = 254–255 region that is to be ascribed to a singly-charged negative parent ion of the benzo[cd]pyrene-6-one molecule.

This agreement between the calculated and observed isotopic compositions and intensities for the M/z = 254, 254,5 and 255 peaks attributed to the doubly-charged dimer of I causes one to give more credence to Dougherty's claim that these ions are due to a doubly-charged polyatomic negative ion. The lack of any similar observations for the other structurally related molecules (II and III) he studied and the lack of two strongly electronegative or electron-withdrawing groups in the monomer of I (although two carbonyl oxygen atoms would be present in the dimer of I) diminishes this credibility somewhat. As will be discussed below, preliminary theoretical calculations suggest that the $C_8H_8{}^{2-}$ ion may be found to exist in the gas phase.

In summary, Dougherty[94] has observed a peak in the negative ion mass spectrum of benzo[cd]pyrene-6-one that he attributes to the doubly-charged negative ion of the dimer of I. Isotopic abundances appear to agree with this assignment. However, no other information is available to substantiate this assertion or to assist in the characterization of this ion and its formation. A thorough reinvestigation of these three molecules by other techniques and/or instrumentation would be most helpful in assessing Dougherty's observations and conclusion.

As mentioned earlier, Bowie and Blumenthal[45] had searched unsuccessfully for doubly-charged negative ions with a conventional mass spectrometer that employed a Nier-type ion source. Bowie, Hart and Blumenthal[46] also modified this spectrometer to provide for metastable transition and kinetic energy release studies by the electric sector voltage variation technique[168, 249]. Many other studies[297] of negative ion mass spectra have not reported the observation of any Z^{2-}, AZ^{2-} or more complex doubly-charged negative ions. Recently, however, Bowie and Stapleton[47] employed a conventional double-focusing Nier-Johnson geometry mass spectrometer to observe an intensity ratio of the doubly-charged molecular ion to the singly-charged molecular ion, M^{2-}/M^-, of 10^{-3} from p-nitrobenzoic acid at an accelerating potential of 3.6 kV. This and the other doubly-charged negative ions they observed must have lifetimes of at least 10^{-5} s in order to reach the detector.

The methods employed in these studies are basically those of kinetic energy and momentum/charge analyses, as used in metastable transition studies[168, 249]. The approach is similar to that employed by the Heidelberg group, and Eq. (25) through (29) apply here as well. The kinetic energy analysis discussed earlier (Eqs. (33) through (37)) is effected in a somewhat similar manner, but with cylindrical electric sector plates rather than the parallel plates; thus, the ion beam is deflected in the x–y plane rather than in the z-direction. An electron beam of nominally 70 eV was used and the pressure in the first field-free region between the acceleration field and the electric sector was approximately 3×10^{-5} Torr.

The $-E/2$ spectra, in which the M^{2-}, $(M-H)^{2-}$, NO_2^{2-} and other Z^{2-} ions were detected by Bowie and Stapleton, are negative ion mass spectra recorded with an electric sector voltage of $1/2$ of that required to transmit ions that have acquired a kinetic energy of T by means of full acceleration through the potential drop V. Therefore, negatively-charged species that have only $1/2$ of the kinetic energy given by Eq. (27) are transmitted through the electric sector. As an example, an NO_2^- (M/z = 46) ion that undergoes an electron attachment after acceleration through V to form an NO_2^{2-} ion (M/z = 23) will have a kinetic energy of $T_a = T/2$, where T is the kinetic energy that a primary NO_2^{2-} ion would have upon achieving full acceleration through V. The apparent M/z of the NO_2^{2-} ion formed by attachment is $23^2/46 = 11.5$ upon momentum/charge analysis in the magnetic sector. This M/z value is observed for the NO_2^{2-} species shown in Bowie and Stapleton's Fig. 1. Disquieting, however, is the apparent mis-labeling of the abscissa for the M^{2-} and $(M-H)^{2-}$ peaks shown in the same figure.

In addition to several doubly-charged polyatomic negative ions, Bowie and Stapleton[47] found Cl^{2-} produced from CH_3Cl, $CHCl_3$ and CCl_4 and Br^{2-} formed from CH_3Br and $CHBr_3$. They also detected I^{2-} formed from CH_3I. These X^{2-} ions were produced in small abundance, whereas the doubly-charged polyatomic negative ions were more intense. Although triatomic NO_2^{2-} ions were formed from the nitrobenzoic acids (and from p-dinitrobenzene, a result not found in the earlier study by Bowie and Stapleton[45]), no indication is given of the formation of any diatomic CN^{2-} ions from the cyanobenzoic acids. However, both the cyano- and the nitro-benzoic acids were found to yield M^{2-} and $(M-H)^{2-}$ ions.

A host of principally aromatic organic molecules that contain one or more electron-withdrawing substituents were scrutinized for doubly-charged negative ion formation. A large fraction of those examined were not found to form either fragment or molecular doubly-charged negative ions. Bowie and Stapleton generalize that benzenoid derivatives with two electron-withdrawing substituents attached directly to the aromatic ring are unable to form doubly-charged molecular anions; a few exceptions to this (p-dinitrobenzene and the cyano- and nitro-benzoic acids) are noted.

p-Nitrophenyl-butanoic and -pentanoic acids, and their methyl esters, were found to provide the largest abundances of M^{2-} ions in Bowie and Stapleton's study. From a study of the analogous 4-(p-nitrophenyl) cyclohexylcarboxylic acid (and ester, they concluded

(1) that a stabilization occurs in the first-mentioned compounds "because suitable coiling of the side chain enables the charged centers to interact", and

(2) that, in general, the two charged sites should be at least 7–8 A apart.

In the various polyatomic molecules they examined, M^{2-} and $(M-H)^{2-}$ ions were produced from acids and esters, but only $(M-CH_3)^{2-}$ ions were formed from the ethers.

A pressure study was made by Bowie and Stapleton to learn something of the mechanism of formation of the doubly-charged negative ions. They considered the collision-induced reaction

$$Z^- + M \longrightarrow Z^{2-} + M^+ \tag{38}$$

and the simple electron capture process

$$Z^- + e \longrightarrow Z^{2-} \tag{39}$$

and concluded that since they found the Z^{2-} current to increase as the sample pressure of Z was increased in the source, the low energy electron capture process, thought to occur just inside the entrance to the electric sector, was the more plausible explanation. A maximum intensity of Z^{2-} was located at about 4×10^{-5} Torr, and at higher pressures of Z the current of Z^{2-} was found to decrease; this decrease they ascribed to collisional scattering. Moreover, addition of a collision gas, M, into the first field-free region (after the acceleration region and prior to the electric sector) caused a decrease by about a factor of two in the intensity of Z^{2-}. This decrease was not strongly dependent upon the nature of M (He, Ar and toluene were employed).

This author has some difficulty with the mechanism for formation of the doubly-charged negative ions suggested by Bowie and Stapleton. Stuckey and Kiser previously discussed and eliminated the simple electron capture process (compare Reactions (5), (17) and (39)) that involves Z^- in the ground state. The suggestion by Bowie and Stapleton that the source of secondary electrons for capture via Reaction (39) is due to Z^- ions striking the sector plate is open to question. If the ion beam current of Z^- was as great as 10^{-10} A in their experiments, *very low* currents of the proposed slow electrons would be formed (only about 10^{-6} of the electron beam current). To obtain doubly-charged negative ion currents six orders of magnitude less than that of the Z^- species would require a very large capture cross section. And the reason for a sector voltage of $-0.506E$ rather than the expected $-0.500E$ for transmission of the Z^{2-} ions is unclear, particularly since their calibration is stated to be $\pm 0.0002E$. Therefore, the conclusion drawn, that this establishes that no translational energy gain is involved, is open to some question.

It is difficult to interpret the experimental findings of Bowie and Stapleton as other than the detection of Z^{2-} species. However, the mechanism of the formation of the doubly-charged ions requires further study and testing in the opinion of this author. The unsuccessful search for X^{2-} ions reported by Chupka, Spence and Stevens is in disagreement with the observations of the X^{2-} ions with an electron bombardment ion source by Bowie and Stapleton. Since so little experimental information and results were provided by Chupka, Spence and Stevens, the extent of this disagreement is uncertain. Additional experiments should assist in clarifying these points.

Dudley, Cady and Crittenden[96] studied the negative ion mass spectra of a variety of fluorine-containing molecules using a double-focusing mass spectrometer of Nier-Johnson geometry and found that in all of the molecules they investigated there were fewer, and less intense, peaks than in the positive ion spectra of those compounds. Further, they noted that the negative ion mass spectra were inferior to the positive ion mass spectra in yielding parent molecular ions for the compounds examined. However, of interest to this review, Dudley, Cady and Crittenden list in their Table 4 a relative abundance of 1.8 % for a species formed from N_2F_4 that must be interpreted (1) as NF_2^{2-}, or (2) as due to an impurity, or (3) possibly as a typographical error.

A careful reading of Dudley, Cady and Crittenden's report indicates no comments by the authors concerning this ion. For each of the other molecules summarized in this table, the positive ion spectrum contains the AF_2^{2+} species. Although it is possible that the 1.8 % belongs in the previous column which lists the intensities of the positive ions formed from tetrafluorohydrazine, the mass spectra of N_2F_4 reported by Herron and Dibeler[137], by Loughran and Mader[186], and by Colburn and Kennedy[64] contain no evidence for the presence of NF_2^{2+} formed from N_2F_4. Thus, the typographical error interpretation is not considered very likely.

Dillard[90] has postulated that this $M/z = 26$ species may be due to CN^- produced from an impurity in the sample used. No detail concerning the source or preparation of the N_2F_4 is given, and therefore Dillard's suggestion remains viable.

The species NF_2^{2-} would necessarily have at least one electron in an antibonding orbital and this, together with a Coulombic repulsion of about 10 eV (assuming the charge separation is about equal to the N–F bond length) and an N–F bond energy of approximately 2.5 eV does not add strength to the interpretation of the $M/z = 26$ ion as due to NF_2^{2-}.

This author concludes that at the present time there is insufficient evidence to argue for the existence of an NF_2^{2-} species. However, its similarity to the NO_2^{2-} ion and the presence of the highly electronegative F atom suggest that additional experiments be conducted in a effort to confirm the existence of an NF_2^{2-} ion.

D. H^{2-} Formed in a Duoplasmatron Ion Source

The existence of the H^{2-} ion has been a controversial topic. Levy-Leblond[179] has "proven" on the basis of a geometrical consequence of the Coulomb Law that any system of four particles with three equal and one opposite charge, regardless of the masses involved, has no stable bound state. That is, the H^{2-} ion, consisting of a proton and three electrons, cannot exist. Avron, Herbst and Simon[11] clearly take issue with the Levy-Leblond statement when they say that there is no rigorous proof that the H^{2-} ion has no binding. The existence of the H^{2-} ion is inferred from several experimental observations.

Isoelectronic with He^-, H^{2-} ions apparently have been formed in crossed ion and electron beams. Walton, Peart and Dolder[302, 303] measured the cross section for H^- detachment using inclined beams of electrons and ions. They found a pronounced structure in the detachment cross section at an energy of ca. 14.2 eV above the ground state of H^- that they attributed to the formation of a resonant state of H^{2-} with a lifetime of about 10^{-15} s. The lifetime estimate was obtained using the uncertainty relation and was based on an observed structure width of approximately 1 eV.

Taylor and Thomas[285] considered this resonance where an electron is absorbed into H^- to form H^{2-}. Using a stabilization technique, Taylor and Thomas found that their calculations for the $2s^2 2p^1 \, ^2P^0$ state of H^{2-} agreed well with the experimental results of Walton, Peart and Dolder. Further, the calculations also showed a possibility of another resonance at a higher energy with a largely $2p^3$ wave function. In additional experiments, Peart and Dolder[232] confirmed the structure they found earlier at about 14.2 eV and found, in addition, the predicted second resonance at an energy of ca. 17.2 eV relative to the H^- ground state. This latter resonance at 17.2 eV they suggested may be ascribed to the possible $2p^3 \, ^2P^0$ state of H^{2-} noted by Taylor and Thomas. Thomas[286] has since calculated that a state that is about 60 % $2p^3$ and about 20 % $2s^2 2p^1$ exists with an energy of 17.26 eV. Dolder and Peart[92] also recently reviewed these results.

Both of these resonances are very short-lived. A much longer-lived H^{2-} species has been reported by Schnitzer and Anbar[10, 255, 256]. Much as with the short-lived H^{2-} ion, the longer-lived H^{2-} ion was not detected as such, but was inferred from the experimental data. The work of Schnitzer and Anbar has been criticized[1, 296] and additional experiments suggested. An assignment of the species studied by Schnitzer and Anbar has been suggested[97] and responses to the criticisms that include additional experimental findings have been offered by Schnitzer and Anbar[256].

Schnitzer and Anbar's experiments are rather similar to those of Baumann, Heinicke, Kaiser and Bethge[26]. Both employed plasma ion sources; that employed by Schnitzer and Anbar was a hollow cathode duoplasmatron. Also, both used einzel lenses and momentum/charge analysis. But the different lifetimes of the doubly-charged negative ions studied dictated somewhat different analyses. Baumann, et al. used an electric deflection analysis after the magnetic sector, as already seen, whereas Schnitzer and Anbar employed a Wien velocity filter and einzel lens voltage variation prior to the magnetic sector.

Extraction and acceleration of the negative ions from the duoplasmatron was accomplished with voltages of 5 to 15 kV. After passing through the einzel lens, the ions encounter a 30 cm Wien velocity filter. The ExB field of this filter allows analysis based only on the ion velocity according to

$$qE = Bev \tag{40}$$

so that ions of velocity

$$v = E/B \tag{41}$$

pass through the filter undeflected. The 90 ° magnetic sector between the detector and the Wien filter then provides the momentum/charge analysis, as indicated in Eqs. (25) to (29).

Schnitzer and Anbar[255] conducted their experiments using both hydrogen and deuterium. We shall consider only the hydrogen case here, since the results of the deuterium experiments duplicated those of the hydrogen experiments. A primary H^{2-} ion accelerated from the source through a potential drop of V will acquire a velocity that is twice as great as a primary H_2^- ion and $2^{0.5}$ that of a primary H^- ion both accelerated through the same potential drop. After passage through the Wien filter and the magnetic sector, the primary H^{2-} ion is detected at $M/z = 1/2$ whereas the primary H_2^- ion is detected at $M/z = 2$ and a primary H^- ion is detected at $M/z = 1$. Since the einzel lens acts as a low resolution kinetic energy/charge analyzer,

$$T = qV_e \tag{42}$$

and both of these ions have acquired the same kinetic energy, both ions will be transmitted at the same value of the einzel lens voltage, V_e.

Now consider an H_2^- ion that undergoes a metastable transition to form a secondary H^- ion, H_s^-

$$H_2^- \longrightarrow H_s^- + H \tag{43}$$

and an H^{2-} ion that undergoes autodetachment to form a secondary H^- ion, H_a^-

$$H^{2-} \longrightarrow H_a^- + e \tag{44}$$

after full acceleration but prior to passage through the einzel lens. The velocity of H_s^- will be twice as great as that of H_a^-. The apparent M/z at the Wien filter for both ions will be as above, but the apparent M/z of the H_s^- ion in the momentum/charge analysis will be 1/2 whereas the apparent M/z of the H_a^- ion will be 2. And the einzel lens voltage must be twice as great to transmit the H_a^- ion as that required to transmit the H_s^- ion.

By varying the extraction voltages on the skimmer, einzel lens and Wien filter, the ion currents due to the various species were detected and the H_a^- ion due to Reaction (44) was identified. Thereby, the existence of H^{2-} was inferred by Schnitzer and Anbar.

Collisional detachment, i. e.,

$$H^{2-} \longrightarrow H + 2e \tag{45}$$

was found to be a relatively unimportant decay channel of the H^{2-} ion. With the assumption that the autodetachment process is a first order decay process for H^{2-}, changes in the potential on the skimmer permitted kinetic studies from which Schnitzer and Anbar found the half-life on the H^{2-} species to be 23 ± 4 ns. Also, they detected some field-induced detachment by varying the potentials on the einzel lens elements.

The results were duplicated when D_2 was employed in place of H_2 in the duoplasmatron; only the voltages and apparent M/z values were altered, and a D^{2-} ion with essentially the same relatively long lifetime was inferred from the results.

Pressure studies indicated a second power dependence of the H_a^- ion current on the pressure in the einzel lens chamber. Since similar effects were observed with Ar, Ne, H_2, Kr and Xe, Schnitzer and Anbar conclude that the effect is not one solely involving H_2. They concluded from these studies that the collisional effect is relatively insignificant and that the primary effect of the pressure is to enhance the initial production of H^{2-} ions and/or their extraction from the source.

A three-step mechanism for the production of H^{2-} was proposed by Schnitzer and Anbar that is consistent with their experimental findings:

$$Q + e \longrightarrow Q^+ + 2e \tag{46}$$

$$Q^+ + H^- \longrightarrow H^{-*} + Q^+ \tag{47}$$

$$H^{-*} + e + Q \longrightarrow H^{2-} + Q \tag{48}$$

where Q is a neutral gas.

That is, a fast Q^+ ion, formed by electron impact and accelerated toward the plasma surface, excites the H^- ion to form H^{-*}, and then the H^{-*} ion reacts via a third body electron capture process to form the H^{2-} ion inferred from the experiments. Only the H^{2-} ions formed at the surface of the plasma could be accelerated through the same potential drop as the primary H^- ions. Further, Schnitzer and Anbar argue that since there is no significant change in the H_a^- ion current as the ion source pressure is varied, the formation of H^{2-} does not occur within the plasma; rather, they conclude that the formation of H^{2-} occurs at the plasma surface.

The states of H^-, H^{-*} and H^{2-} were not specified by Schnitzer and Anbar. They do note that both the H^{2-} and the H^- precursor must be highly excited, a condition that can be achieved by the ion-ion collision in the second step of their mechanism [Reaction (47)]. The efficiency for the ion-ion collision is expected to be relatively high because of the cross section for this process. Thus, Ar should be better than He. However, since Ar is so much heavier than H, Schnitzer and Anbar argue Ar will be less efficient; this argument is necessary in order to explain the observation that the H_a^- ion current exhibited less than a second order dependency on the argon pressure in the lens chamber.

Durup[97] has provided an electron configuration assignment for the H^{2-} species with a lifetime of 23 ns as reported by Schnitzer and Anbar. He suggests that this is the $2p^3 \, {}^4S^0$ state of H^{2-}, a state that is symmetry-forbidden both with respect to decay to the $1s^2 \, H^-$ state [Reaction (44)] and with respect to decay to $1s^1 \, H$ [Reaction (45)]. Further, Durup notes that this state could radiate only through a dielectronic transition to the electronic double continuum. Durup indicates that the $2p^3 \, H^{2-}$ ion configuration may be viewed as purely $2p_x 2p_y 2p_z$ and there would be less Coulombic repulsion between electron pairs in this than any other configuration involving 2p electrons. He finds that the metastability of the $2p^3 \, {}^4S^0$ state of H^{2-} would mean that the energy of this state is lower than that of the $2p^1 \, H$ atom, and that this H atom could bind two other 2p electrons.

Vestal's[296] criticism of Schnitzer and Anbar's interpretations centered on the possibility that double charge exchange processes might occur in the einzel lens. Vestal suggested that experiments using other than the discrete energies of 2V, V and V/2 be employed to test for collisional processes occurring in the drift region and einzel lens. Also suggested was a variation of the voltage rather than use of discrete settings of the Wien filter. Schnitzer and Anbar[256] performed these additional experiments and concluded, based on the experimental results they obtained, that the charge transfer process did not account for their observations. The criticisms of Aberth[1], more extensive than those of Vestal, were concerned with many of the same points and with the importance of the pressure effects. Schnitzer and Anbar's[256] additional experiments and findings on the pressure studies, included in the above discussions, reportedly excluded the double collisional processes as significant contributors to the observed H^- ion currents.

It appears that doubt still lingers concerning the existence of a long-lived H^{2-} ion. It is of particular importance that other investigators repeat these experimental studies and report their findings. There are many similarities in the experiments of Baumann, Heinicke, Kaiser and Bethge and the experiments of Schnitzer and Anbar. It has been reported[153] that Feldmann has been unable to repeat the formation of doubly-charged negative ions with the same apparatus used by Baumann, Heinicke, Kaiser and Bethge, and serious questions have been raised[1, 296] about the interpretations given by Schnitzer and Anbar to their experimental observations. Indeed, if instrumental artifacts associated with the use of the einzel lenses are the cause of the observations of Schnitzer and Anbar, these should be experimentally determined. Independent experimental verification both of Schnitzer and Anbar's observations and of the findings of Baumann, Heinicke, Kaiser and Bethge would significantly diminish the doubts expressed.

The ICR experiments of Stapleton and Bowie have verified many of the observations of Stuckey and Kiser. Additionally, several groups have reported doubly-charged negative ions formed either in the ion source or in the accelerating region or in the electric sector of various types of mass spectrometers that used electron bombardment ion sources. Therefore it is now fitting to examine the possible states of these ions and to discuss mechanisms which might exist for the formation of the doubly-charged negative ions.

V Electronic States of Doubly-Charged Negative Ions

At present no experimental data are available that give direct information concerning the electronic states of the doubly-charged negative monatomic or molecular ions that have been observed. However, data are available that may be utilized by extrapolation and applied to the consideration of a few of the doubly-charged negative monatomic ions. The electronic structures of the diatomic and polyatomic AZ^{2-} ions may be considered by employing quantum mechanical calculations. Both of these approaches shall be adopted in this attempt to understand more fully the doubly-charged negative ions.

Many types of extrapolations of spectroscopic data for isoelectronic species have been proposed previously in the attempt to obtain electron affinities for a variety of species. In the present efforts, the equation offered by Glockler[116] is used in conjunction with term differences for a variety of spectroscopic states in isoelectronic sequences.

To obtain some information about the electronic states of the molecular species, calculations with the extended Hückel theory (EHT) have been made for several of these molecular ions. Lowe[187] has applied qualitative molecular orbital concepts to the electron affinities of a number of polyatomics and Younkin, Smith and Compton[316] have employed semiempirical methods in considering the electron affinities of a group of conjugated organic molecules. In this work, the Hoffmann[146–149] EHT treatment is employed; VSIP values obtained from a Hartree-Fock-Slater approximation are used and the Mulliken-Wolfsberg-Helmholtz arithmetic mean is chosen for the Hamiltonian construction. Although it is recognized that the EHT calculations have certain limitations, the use of the EHT approach is justified by the ease and speed of the computations to provide zero *th* order approximations and some additional insight, particularly in view of the lack of any other information.

Since some guidance in the EHT molecular calculations and interpretations is afforded by an understanding of the atomic systems, the monatomic Z^{2-} species shall be examined first.

A. Monatomic Z^{2-} Species

Until the fairly recent development of photodetachment and other techniques, the electron affinities of atoms were usually only poorly determined from electron impact studies. Consequently, numerous attempts were made to employ a variety of extrapolation procedures with data from within a given isoelectronic sequence in an attempt to obtain more reliable values or estimates where no experimental data existed.

Since there is no direct information yet available for the doubly-charged negative ions, it is necessary at this time to use similar extrapolation techniques. Bartlett[16] and Braunbek[49] employed a linear extrapolation, but Glockler[116] provided a very useful empirical approach that employs a simple quadratic equation. Other methods of estimating electron affinities have been reported by Allred[6], Bates[20], Bates and Moiseiwitsch[21], Baughan[22–24], Crossley and Coulson[70, 71], Edlen[99, 100], Geltman[114], Ginsberg and Miller[115], Glockler[117], Glockler and Sausville[118], Iczkowski and Margrave[156], Johnson and Rohrlich[160, 161], Moiseiwitsch[210], Sheehan[261], and Wu[314]. Zollweg[318] has commented upon the utility of many of these extrapolation techniques.

It has been indicated by Bates and Moiseiwitsch[21, 212] that the original Glockler equation is rather satisfactory when considering the energy difference of two bound terms, as suggested earlier by Bates[20]. The Glockler equation[116] represents the energy as a quadratic function of the charge, z,

$$E = az^2 + bz + c \tag{49}$$

and may be written for the energies of term differences for the first three members of an isoelectronic sequence to evaluate the constants a, b and c, i. e.,

$$\begin{vmatrix} 2^2 & 2^1 & 2^0 \\ 3^2 & 3^1 & 3^0 \\ 4^2 & 4^1 & 4^0 \end{vmatrix} = \begin{vmatrix} E(2) \\ E(3) \\ E(4) \end{vmatrix} \qquad (50)$$

From Eq. (50) it is found that

$$a = [E(2) - 2E(3) + E(4)]/2$$
$$b = -[7E(2) - 12E(3) + 5E(4)]/2 \qquad (51)$$
$$c = 6E(2) - 8E(3) + 3E(4)$$

Therefore, for extrapolations to singly-charged negative species

$$E(1) = 3E(2) - 3E(3) + E(4) \qquad (52)$$

and for two-step extrapolations to doubly-charged negative ionic species

$$E(O) = 6E(2) - 8E(3) + 3E(4) \qquad (53)$$

The source of spectroscopic data for the various electronic states of the different neutral and positive ion species in isoelectronic sequences uniformly has been the

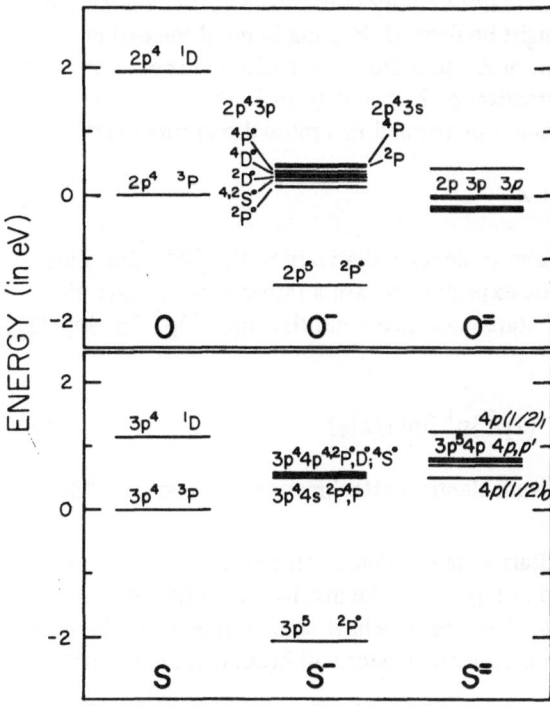

Fig. 19. Selected energy levels in the oxygen and sulfur systems, as determined from isoelectronic sequences

tabulation by Moore[213]. With these data, extrapolations have been made with the appropriate isoelectronic sequence to give the results indicated in Fig. 19 for O^- and O^{2-} and S^- and S^{2-}. The electron affinities (see Table 1) for O and S and the excited states[213] of O and S are experimental values; all other values were obtained using Eqs. (52) and (53).

Two important points are to be noted. First, both O^{2-} and S^{2-} have one or more states that are estimated by Eq. (53) to lie below the np^4 $(n+1)s$ and np^4 $(n+1)p$ manifolds for O^- and S^-, respectively. Second, the np^4 $(n+1)s$ and np^4 $(n+1)p$ manifolds for O^- contain states that are calculated from Eq. (52) to lie only slightly above the ground state of the respective neutral atoms. This is just as suggested earlier by Bates[20] and employed by Bates and Moiseiwitsch[21]. In the absence of any experimental data, it is not unreasonable to suggest that one or more states of the respective manifold may actually exist below the ^3P ground states of O and S. For purposes of discussion, it is supposed that the np^4 $(n+1)p$ $^4P^0$ is just such a state.

The formation of O^{2-} might then proceed as follows. A low energy electron attaches to the ^3P ground state of the oxygen atom to form $O^-(2p^5$ $^2P^0)$.

$$O\,(2p^4\,^3P) + e \longrightarrow O^-\,(2p^5\,^2P^0) \tag{54}$$

This ion may be excited to the $^4P^0$ state of O^-,

$$O^-\,(2p^5\,^2P^0) + Q \longrightarrow O^{-*}\,(2p^4 3p^1\,^4P^0) + Q \tag{55}$$

Of course, the states suggested here are not be only ones possible; for example, in Reaction (55) a $2p^4 3p^1$ ^4S state might be formed. Bearing in mind the earlier comments concerning the formation of Z^- ions from AZ molecular species and the difficulties with the subsequent formation of Z^{-*} ions from Z^- by a neutral Q species in Reaction (55), the O^{-*} might be formed in a more direct process such as

$$OZ + e \longrightarrow O^{-*}\,(2p^4 3p^1\,^4P^0) + Z \tag{56}$$

The product O^{-*} is LaPorte-forbidden to decay radiatively to the $^2P^0$ state. This metastable $^4P^0$ state of O^- would be expected to have a rather long lifetime. Electron attachment to the excited $^4P^0$ state may then yield the triplet $2p^5$ $3p'$ $3p[1/2]_0$ state of O^{2-}.

$$O^{-*}\,(2p^4 3p^1\,^4P^0) + e \longrightarrow O^{2-}\,(2p^5 3p^1\,3p[1/2]_0) \tag{57}$$

If a $2p^4 3s^1$ ^4S state of O^{-*} is involved, electron attachment may yield a $2\,p^4 3s^1 3p^1$ $^5S^0$ state of O^{2-}.

Excited states involving 3d orbitals were considered also in these extrapolations. It was found that the 3d orbitals do not appear to be involved with the lighter elements that were treated, although they begin to have some influence in the phosphorus system. Presumably any d orbital involvement will become more important with the heavier elements.

Interestingly, extrapolations by this author with the Glockler equation using Ar(I), K(II) and Ca(III) excited state data[213] yields the prediction of doubly-excited Feshbach states of Cl^- at about 8.5, 8.6, 9.9, 10.0 and 11.4 eV relative to Cl. These extrapolated results are in good agreement with energies calculated by Matese, Rountree and Henry[198] using configuration interaction, and with the experimental results of Cuningham and Edwards[72] who collided Cl^- ions extracted from a duoplasmatron with hydrogen and helium gas targets.

An identical sequence of reactions may be proposed for the S^{2-} formation:

$$S\,(3p^4\,{}^3P) + e \longrightarrow S^-\,(3p^5\,{}^2P^0) \tag{58}$$

$$S^-\,(3p^5\,{}^2P^0) + M \longrightarrow S^{-*}\,(3p^4 4p^1\,{}^4P^0) + M \tag{59}$$

$$S^{-*}\,(3p^4 4p^1\,{}^4P^0) + e \longrightarrow S^{2-}\,(3p^5 4p^1\,4p[1/2]_0) \tag{60}$$

with the analogous possibility for the more direct formation of S^{-*} according to

$$SZ + e \longrightarrow S^{-*}\,(3p^4 4p^1\,{}^4P^0) + Z \tag{61}$$

Of course, Reactions (54) and (58) may be replaced by reactions that involve an appropriate dissociative resonant capture to yield the same ground state of O^- or S^-.

It is important to recognize here that the excited states indicated are specified only for convenience in this discussion. They are plausible, but there is no direct evidence that these are the specific excited states involved in the formation of the doubly-charged negative ions. Further experiments will be necessary to establish the electronic configurations of these excited states.

The above mechanisms for the formation of O^{2-} and S^{2-} are plausible and explain the observation of O^{2-} and S^{2-} by Stuckey and Kiser and by Baumann, Heinicke, Kaiser and Bethge. They specifically involve excited states both of Z^- and of Z^{2-}. And the lifetime of the Z^{-*} would permit significant concentrations of Z^- to be formed in a plasma source, such as found in the omegatron or the Penning source. However, it is anticipated that only much lower concentrations of Z^{-*} would be formed in the Nier-type ion source employed in most conventional mass spectrometers. It is unlikely, therefore, that the above mechanisms can satisfactorily account for the observation of F^{2-} formed from CF_3Cl, as reported by Ahnell and Koski[4], or the various Z^{2-} species reported by Bowie and Stapleton[47].

One might wish to continue these isoelectronic sequence extrapolations to the cases of Se^{2-} and Te^{2-}, species observed by Baumann, et al. However, the absence of sufficient spectroscopic data does not permit such calculations at this time. Therefore, the arguments for these species must be based on analogy and chemical similarity to the earlier chalcogens. Possible d orbital involvement is unknown at this time, and therefore simply has been neglected.

Analogous extrapolations are, however, possible for the P^- and P^{2-} ions. The results are quite closely related to those found for the O and S cases discussed above. P^- is estimated to have a manifold of excited states about 1.1 eV above the ${}^3P$ state, even as noted by Bates and Moiseiwitsch[21]; these states are the $3p^3 4p^1$ $^3P,^5P$ and the $3p^3 4s^1$ $^3S^0, ^5S^0$. In the calculations for P^{2-}, which involve a two-step extrapolation, it is found that a manifold of $3p^4 4p^1$ states occur 0.4 to 0.9 eV *above* the center of the P^{-*} $3p^3 4p^1/3p^3 4s^1$ manifold. The lowest of these P^{2-} states is taken to be the $^4S^0$, although the extrapolations indicate that the ^2D state of P^{2-}, a $3p^4$ (^3P) $3d^1$ configuration, may lie nearly 1.5 eV lower than the $^4S^0$ state.

It is considered reasonable to utilize the same mechanism for P^{2-} formation; for example,

$$P (3p^3 \ ^4S) + e \longrightarrow P^- (3p^4 \ ^3P) \tag{62}$$

$$\text{and} \quad P^- (3p^4 \ ^3P) + M \longrightarrow P^{-*} (3p^3 4p^1 \ ^5P) \tag{63}$$

$$\text{or} \quad PZ + e \longrightarrow P^{-*} (3p^3 3p^1 \ ^5P) \tag{64}$$

$$P^{-*} (3p^3 4p^1 \ ^5P) + e \longrightarrow P^{2-} (3p^4 4p^1 \ ^4S^0) \tag{65}$$

Extension of these processes to similar mechanisms for the As^{2-} ion and its cognates would satisfactorily explain the formation of P^{2-}, As^{2-}, Sb^{2-} and Bi^{2-} observed by Baumann, Heinicke, Kaiser and Bethge[26].

In the suggested processes for the formation of Z^{2-} species, the author has attempted to indicate that the Z^{2-} species may be bound. Experimental verification of this stability of the Z^{2-} ions is lacking at this time. However, it must be recognized also that the energy requirement for the Z^{2-} species is not as important as is the lifetime of the Z^{2-} species. If the Z^{2-} ion is formed in an excited state of high multiplicity that has a sufficiently long lifetime against decay through the various channels possible, then the Z^{2-} species may be observed even though it is not a bound state. Nazaroff[221] has commented that as long as the negative ion does not become "too unbound", one can talk in terms of a negative ion; how much is "too unbound" is not specified.

The doubly-charged negative halogen ions remain to be considered. Unfortunately, there are insufficient spectroscopic data available at this time to treat the doubly-charged negative ions of any of the halogens. Therefore, resort to analogy again must be sought. With He^{-*} as a model, and considering the above comments on the O^{2-}, S^{2-} and P^{2-} ions, reasonable interpolation provides some enlightenment concerning the properties of the doubly-charged negative ions of Group VII.

It would appear that since He^{-*} is just bound, the heavier and more polarizable noble gas atoms also may be bound. Indications given above suggest that O^{2-}, S^{2-} and P^{2-} may be bound or nearly so. Interpolations to F^{2-} and Cl^{2-} then would indicate that the X^{2-} species also may be bound or nearly so. However, as noted previously, it is more important that the X^{2-} species be formed in excited states of high multiplicity to provide for long lifetimes of these ions. For this reason, and by analogy to the He^{-*} model, the observed X^{2-} ions are not expected to be found in the np^6 $(n+1)s^1$ ^1S states. Rather, the X^{2-} species may have electronic configurations such as np^5 $(n+1)s^1$ $(n+1)p^1$ ^4D, or even exist in sextet states as suggested by Massey[196].

B. Diatomic CN^{2-} and Triatomic NO_2^{2-} Species

The excited states of doubly-charged negative ions larger than monatomic are also of importance in providing for long lifetimes of these species. Studies of the energy stabilities of di-, tri-, and poly-atomic Z^{2-} ions involve molecular orbital treatments, however, rather than the approaches just discussed for atomic Z^{2-} species. A variety of semiempirical molecular calculation methods are possible.

The extended Hückel theory calculations, used in this work and discussed below, are based on the approaches of Hoffmann[146−149]. Although VSIP values given by Cusachs, Reynolds and Barnard[76, 77] were explored for use as the Coulomb integrals, the VSIP values obtained from a Hartree-Fock-Slater approximation by Herman and Skillman[135] were consistently used in the present EHT calculations by this author. Both the geometric mean formula due to Mulliken[217, 218] and Cusachs formula[74] were considered for the Hamiltonian construction, but the Mulliken-Wolfsberg-Helmholtz arithmetic mean formula[217, 218, 313] was chosen for use. Cusachs and Corrington[75] have compared various parametrizations for semi-empirical molecular calculations.

The basic overlap integrals are calculated from the formulas given by Mulliken, Rieke, Orloff and Orloff[219]; d orbital overlap integral formulas are those given by Jaffe and Doak[157, 158]. (Two-center overlap integral formulas are given by Lofthus[184].) Slater[267] exponents were employed. 1.875 was taken as the constant rather than 1.75 in constructing the resonance integrals. No other scaling was used in the calculations.

As is well known, geometries found with EHT calculations are rather good. Although binding energies are somewhat less satisfactory, in this work it was determined that in the vicinity of the lowest unoccupied molecular orbital (LUMO) and the highest occupied molecular orbital (HOMO), the energies were reasonably good and excitation energies (vertical and adiabatic) to the lower electronic states were generally in good agreement with available experimental data. However, charge distribution were poorly estimated, a general failing of the EHT approach.

The ground state configuration of the CN molecule is $(3\sigma)^2 (4\sigma)^2 (1\pi)^4 (5\sigma)^1$ $^2\Sigma^+$ (see Fig. 20). Schaefer and Heil[252] have discussed the electronic structures of the low-lying states of the CN molecule. The HOMO can accommodate another electron to form the CN^- ion. CN, a pseudo-halogen, has an electron affinity that is among the highest known (see Table 1). The formation of CN^- by placing an electron in the 5σ molecular orbital leads to an increased C–N bond strength and a shortening of the C–N bond length. Addition of another electron to form CN^{2-} might then lead to a configuration of $(3\sigma)^2 (4\sigma)^2 (1\pi)^4 (5\sigma)^2 (2\pi)^1$ $^2\Pi$, since the electron would enter the antibonding 2π molecular orbital. The 2π virtual level is found from the EHT calculations to lie only moderately higher than the 5σ orbital. Dogett and McKendrick[91] have used LCAO-MO-SCF calculations with a Clementi double-zeta basis set and find an electron affinity of 3.3 eV with Koopmans' theorem[174] for $^2\Sigma^+$ CN to form the $^1\Sigma^+ CN^-$ ion.

In view of the electronic configurations of the monatomic species it appears reasonable to believe that the electronic configuration of CN^{2-} is not that indicated just above, but rather that it is $(3\sigma)^2 (4\sigma)^2 (1\pi)^4 (5\sigma)^1 (2\pi)^2$, i. e., a $^4\Sigma$ state.

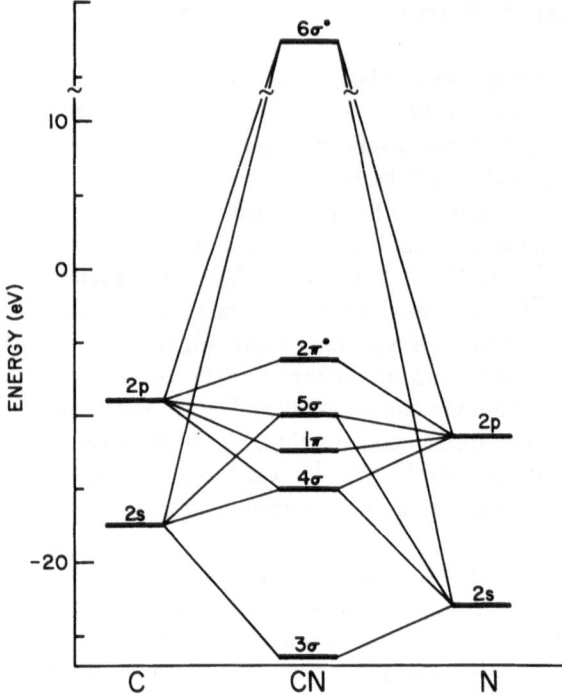

Fig. 20. Molecular orbital diagram for CN based upon extended Hückel theory calculations

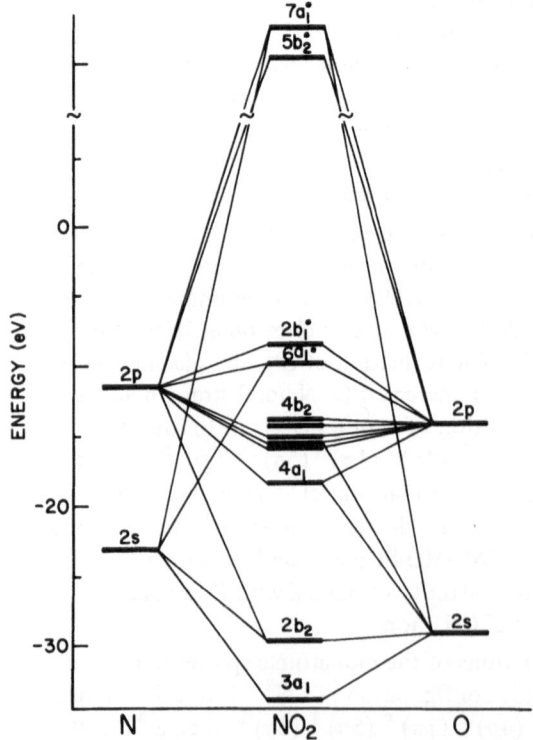

Fig. 21. Molecular orbital diagram for NO_2 based upon extended Hückel theory calculations

The CN^{2-} ion could be formed in the omegatron source by excitation of the CN^- $^1\Sigma^+$ ground state to the $^3\Pi$ state, and this CN^{-*} could capture another electron to form the $^4\Sigma$ state (or, of course, CN^{2-} could be formed from the $^3\Pi$ CN^{-*} state produced directly from CF_3CN by the analogous direct dissociative capture process). The 5σ orbital pairing energy would be saved and the half-filled 2π orbital could also assist in decreasing the overall energy requirements. The diffuseness of the 2π orbital also assists in reducing Coulombic repulsion. Thus, it is plausible that the CN^{2-} ion exists in a quartet state, rather analogous also to the model He^{-*} ion with the configuration $1s^1 2s^1 2p^1$ ^4P. As a result of the high multiplicity, the CN^{2-} ion may be expected to have a fairly long lifetime against electron detachment.

The molecular orbital diagram for NO_2 is presented in Fig. 21. The ground state configuration of the NO_2 molecule, $(3a_1)^2 (2b_2)^2 (4a_1)^2$ $(1b_1)^2 (5a_1)^2 (3b_2)^2$ $(1a_2)^2 (4b_2)^2$ $(6a_1)^1$, is $^2A_1$. The HOMO, $6a_1$, is essentially antibonding in nature. This orbital can accomodate one additional electron to form the NO_2^- ion in the $^1A_1$ state. An excited state of $^3B_1$ lies above the $^1A_1$ state by about 3.5 eV. The next electron to be added, to form NO_2^{2-}, may enter the rather high-lying antibonding orbital indicated as $5b_2$. It might be expected that the ground state configuration of the NO_2^{2-} ion would be $(3a_1)^2 (2b_2)^2 (4a_1)^2$ $(1b_1)^2 (5a_1)^2 (3b_2)^2$ $(1a_2)^2 (4b_2)^2 (6a_1)^2$ $(2b_1)^1$, a $^2B_1$ state. However, as suggested, the excited state $... (1a_2)^2 (4b_2)^2 (6a_1)^1 (2b_1)^1 (5b_2)^1$ $^4A_2$ may be formed rather than the ground state. The $5b_2$ orbital shown in Fig. 21 for NO_2 appears to be at a very high energy, but the EHT calculations indicate the energy of this orbital is drastically lowered as the ONO bond angle is decreased and the N–O bond is lengthened in forming the NO_2^- and then the NO_2^{2-} ion.

In the EHT calculations for NO_2^+, NO_2, NO_2^- and NO_2^{2-}, the O–N–O bond angles were 180, 134.25, 115.4 and 111 °, and the bond lengths were 1.154, 1.197, 1.236 and 1.36 Å, respectively. An adiabatic $I(NO_2) = 8.54$ eV is calculated from the total energies and a vertical $I(NO_2) = 9.04$ eV is obtained using Koopmans' theorem[174]. The electron affinity of NO_2 is found to be substantially positive whether the $^1A_1$, $^3B_1$ or $^1B_1$ states are involved. This large positive electron affinity agrees qualitatively with the data of Table 1.

The results indicate that the NO_2 molecule captures an electron to form NO_2^-. The NO_2^- ion may then be excited to the metastable $^3B_1$ state. More likely, the NO_2^{-*} ion is formed directly in the electron capture process by NO_2. In either instance, the excited $^3B_1$ state can add another electron to form the $... (6a_1)^1$ $(2b_2)^1 (5b_1)^1$ $^4A_2$ metastable state of NO_2^{2-}. Huber, Cosby, Peterson and Moseley[154] have recently studied the photodetachment and de-excitation of vibrationally excited NO_2^{-*} that survives several hundred collisions with O_2. The electronically excited metastable NO_2^{2-} also is expected to have a relatively long lifetime.

For both the doubly-charged negative diatomic ion, CN^{2-}, and the doubly-charged negative triatomic ion, NO_2^{2-}, the results of the EHT calculations suggest that the mechanism for the formation of these species may be remarkably similar to that proposed for the monatomic species. The molecule first acquires an electron to form the singly-charged negative ion which then may be excited to form X^{-*}; alternatively, the X^{-*} species may be formed in the initial interaction with the electron.

The X^{-*} ion, in turn, can attach another electron to form a metastable X^{2-} ion with a fairly long lifetime.

It is of importance, therefore, to consider whether the principal components of this same mechanism can also applied to doubly-charged negative ions formed from polyatomic molecules.

C. Polyatomic Species

As discussed above, a number of doubly-charged negative polyatomic ions have been reported by several observers. As the molecule becomes quite large, the problems due to Coulombic repulsion are reduced by increased separation of charge. Thus, the more important species to examine here are the smaller polyatomic molecules. In particular, the reported SF_5^{2-} and SF_6^{2-} are deemed interesting by virtue of the fact that both SF_5^- and SF_6^- are well known species in the gas phase.

Ahearn and Hannay[2] studied mass spectrometrically the decay of SF_6^- to SF_5^- and indicated that a long-lived metastable state of SF_6^- was involved. Spence and Schulz[274] considered the effect of temperature on electron attachment of molecules at low (< 0.2 eV) energies. They generalized that for many polyatomics the total cross section for negative ion production tends to an absolute maximum as the gas temperature is increased, in agreement with Fox and Curran[110]. Spence and Schulz also postulated that a common SF_6^{-*} ion is involved in decompositions that yield F^-, SF_5^- and the stable SF_6^-. Henis and Mabie[133] have studied an SF_6^{-*} species formed by capture of very low energy electrons in an ICR spectrometer and found its lifetime to be approximately 0.5 ms.

It might therefore be possible that an electronically excited SF_6^{-*} ion could attach a second electron to form SF_6^{2-}, particularly if a cool plasma ion source were employed. Both Stapleton and Bowie[275] and Stuckey and Kiser have observed the SF_6^{2-} ion with the ICR and omegatron spectrometers.

The configuration for the ground state of SF_6, determined with the EHT treatment without d-orbital inclusion and using an S—F bond length of 1.585 Å and O_h symmetry, is found to be . . . $(1t_{2g})^6 (1t_{2u})^6 (5t_{1u})^6 (1t_{1g})^6 (3e_g)^4$ ^1E_g. This result agrees with the EHT calculations reported by Siegbahn, et al.[264]. Siegbahn and co-workers compare CNDO and EHT calculations in considering the assignments of the photoelectron spectrum of SF_6. Von Niessen, Cederbaum, Diercksen and Hohlneicher[298] have treated the electronic structure of SF_6 in detail. Their work also summarizes many other theoretical studies of SF_6 and provides a recent assignment of the SF_6 photoelectron spectrum. The present EHT treatment indicates that the $6t_{1u}$ virtual level of SF_6 lies very much higher than the $3e_g$ orbital, and one might suspect that it could not capture an electron to form SF_6^-. Yet the existence of an SF_6^- ion is very well established.

If it is assumed that the SF_6^- ion might have C_{2v} (pseudo-D_{5h}) symmetry, with the same S—F bond length and an additional electron occupying an equatorial position, one finds from the EHT calculations that the $16a_1$ orbital arising from the $6t_{1u}$ virtual orbital of octahedral SF_6 has been dramatically lowered in energy. Yet the electron affinity of SF_6 based on Koopmans' theorem[174] is not in close agreement with the experimental value. The resonant capture of SF_6 proceeds at essentially 0 eV.

A similar calculation for the SF_6^{2-} ion in which the S–F bond length is kept constant and C_{2v} (pseudo-D_{5h}) symmetry is assumed, indicates that the energy of the $16a_1$ orbital is decreased further and that the energy of the $8b_1$ orbital arising from the $6t_{1u}$ virtual orbital of octahedral SF_6 remains nearly constant.

As a result of these calculations, it is concluded that the SF_6 molecule will not bind two electrons to form SF_6^{2-}. However, the formation of SF_6^{2-} in an electronically excited state cannot be dismissed easily. Nuclear-excited Feshbach resonances, in which the kinetic energy of the incident electron is absorbed solely into the nuclear motion of the molecule, occur in the electron attachment process of various molecular species[61, 62, 123, 124]. The SF_6^{-*} species studied by Henis and Mabie[133] had a lifetime of about 0.5 ms. A lifetime of 0.03 ms, shorter than the value reported by Henis and Mabie, has been discussed and used for equipment calibrations by Hadjiantoniou, Christophorou and Carter[62, 123, 124]. It is conceivable that nuclear-excited Feshbach resonances also may participate in the formation of the SF_6^{2-} ion and other doubly-charged negative ions.

The SF_5^- ion is readily formed from SF_6 in electron bombardment ion sources. Henis and Mabie[133] found the SF_5^- ion to have an essentially infinite lifetime. With the assumption that SF_5 has a square pyramidal structure in which the S atom lies 0.2 Å above the center of the base, SF_5 has been calculated with the EHT approach to have an electron affinity of about 2.2 eV with the S–F bond length equal to 1.585 Å. This may be compared with the experimental value of 3.66 eV given in Table 1. With this EHT treatment, the electronic configuration for the SF_5^- ion (C_{2v} symmetry) is found to be . . . $(10a_1)^2 (7e)^4 (1b_2)^2 (4b_1)^2 (11a_1)^2$ $^1A_1$.

The SF_5^{2-} ion has been observed both in the ICR spectrometer and in the omegatron. An EHT calculation was made with an S–F bond length of 1.585 Å and pseudo-D_{5h} symmetry for SF_5^{2-}. The five F atoms were viewed as surrounding the S atom in the equatorial plane, with a pair of electrons and a lone electron located in the axial positions. An electronic configuration of . . . $(4e_2')^4 (1a_2'')^2 (3a_2'')^2 (6a_1')^1$ $^2A_1'$ is a plausible structure for the SF_5^{2-} ion. A quartet state of . . . $(4e_2')^4 (1a_2'')^2$ $(3a_2'')^1 (6a_1')^1 (6e_1')^1$ is less favorable on an energetic basis, but could provide for the long lifetime necessary for the observed SF_5^{2-} ion.

Compton[65] has indicated that the electron affinity of UF_6 is $\geqslant 5.1$ eV and that of UF_5 is approximately 4 eV. UF_5^{2-} and UF_6^{2-}, analogous to SF_5^{2-} and SF_6^{2-}, also may exist.

Dougherty[94] searched for the $C_8H_8^{2-}$ ion in the negative ion mass spectrum of cyclooctatetraene, but was unable to detect this ion. The results of EHT calculations suggest that the $C_8H_8^{2-}$ ion may exist and would likely be found in an ICR or omegatron instrument. For the C_8H_8 and $C_8H_8^-$ species, a symmetry of D_{2d} is assumed, whereas a D_{8h} symmetry is assumed for the $C_8H_8^{2-}$ ion. The C–H bond length = 1.09 Å is used throughout. $\angle CCH = 114.450°$, $\angle CCC = 131.1°$, $(C-C)a = 1.379$ Å and $(C-C)b = 1.4425$ Å are employed for $C_8H_8^-$. $\angle CCH = 118.3°$, $\angle CCC = 126.46°$, $(C-C)a = 1.334$ A and $(C-C)b = 1.462$ Å is used for the C_8H_8 molecule. For the $C_8H_8^{2-}$ species, C–C = 1.42 Å and $\angle CCC = 135°$ are employed. The results of the EHT calculations indicate an adiabatic ionization potential of 8.26 eV for C_8H_8; this is close to the experimental photoelectron spectroscopic value of 8.21 eV determined by Dewar, Harget and Haselbach[86]. Dewar, Harget and Haselbach have performed both MINDO/2 and π-approximation calculations for the $C_8H_8^{2-}$ ion.

The $5a_1$ orbital (HOMO) in C_8H_8 is filled and the electron added in forming $C_8H_8^-$ enters in the LUMO designated as $3a_2$. Interestingly, the HOMO of $C_8H_8^{2-}$ is found to be $1e_{2u}$ in the singlet state; however, the LUMO is $1e_{3g}$ and lies only slightly higher in energy than the $1e_{2u}$ orbital. Thus, excited states of $\ldots (3e_{3u})^4$ $(1e_{2u})^3 (1e_{3g})^1$ or $\ldots (3e_{3u})^4 (1e_{2u})^2 (1e_{3g})^2$ for the $C_8H_8^{2-}$ ion are energetically realistic and could provide sufficiently long lifetimes for the $C_8H_8^{2-}$ ion to be observed. To the present, the existence of neither the $C_8H_8^{2-}$ (g) ion nor the $C_5H_5^{2-}$ (g) ion has been reported.

Other small polyatomic system may yield doubly-charged negative ions. The thermochemistry of the O_3^- ion and negative ion-molecule reactions of ozone have been discussed by Lifshitz, Wu, Tiernan and Terwilliger[181]. The present EHT calculations indicate the closed (D_{3h}) form of ozone is less stable than found by Hay, Dunning and Goddard[127], Lucchese and Schaefer[188] and Shih, Buenker and Peyerimhoff[262]. The EHT calculations made here for O_3 indicate that the O_3^{2-} species (assumed $\angle OOO = 109.47°$, C_{2v} symmetry (pseudo-tetrahedral), O–O bond length = 1.278 Å) may be formed in excited states that might provide a rather long lifetime for the O_3^{2-} ion. The EHT calculations of the excited states of ozone approximately agree with those of Messmer and Salahub[208] who used the Xα-SW method. Hay and Goddard[128] and Hay and Dunning[126] also have considered the excited states of ozone; they did not treat the O_3^{2-} ion. To the present, no experimental indication of the existence of O_3^{2-} has been reported. On the basis of the EHT calculations, however, an experimental search for the doubly-charged negative ion of ozone appears warranted.

Another relatively small polyatomic species, the carbonate ion, CO_3^{2-}, is known to exist in calcite in the crystalline state. Cosby and Moseley[68] have reported on the observation of a bound excited state in CO_3^-. This author believes it possible that the CO_3^{2-} and the SO_3^{2-} ions may be found to exist also in the gas phase, although no detailed calculations have been made for either of these species. Although the electron affinity of SO_3 (approximately 1.7 eV) is not as great as that of CO_3, it is rather large and not much different than the electron affinity of the cyclopentadienyl radical. An even greater electron affinity has been found for several of the ZF_n systems, e.g., UF_6 and BF_3.

An interesting result is obtained when the BF_3 system is considered. It was estimated that the electron added to BF_3 would cause the BF_3^- to have a geometry intermediate between the planar BF_3 and tetrahedral. Therefore, the B atom was raised 0.2 Å above the base; all B–F bond lengths were maintained at 1.295 Å as in BF_3. The ground state calculated for this BF_3^- (C_{3v} geometry) with $\angle FBF = 117.66°$ is $^2A_1$. The HOMO is a bonding $6a_1$ molecular orbital of energy = -4.86 eV, and consists mainly of B(2s) and B($2p_z$) atomic orbitals (electron populations of 0.06 and 0.88, respectively). These results are comfortingly similar to the *ab initio* calculations of So[271] for BF_3^- with STO–3G type basis functions.

When BF_3^{2-} is treated similarly, with $\angle FBF = 109.47°$ (i. e., a C_{3v} pseudo-tetrahedral geometry) and a B–F length of 1.295 Å, the results obtained from an EHT calculation indicate that the HOMO consists principally of $B(2s)$ and $B(2p_z)$ atomic orbitals (electron populations of 0.38 and 1.49, respectively). As might have been expected, it is indicated that a "lone pair" of electrons resides on the boron atom in BF_3^{2-}.

From the EHT total energies calculated for BF_3, BF_3^- and BF_3^{2-} (assuming the $1a_1'$, $2a_1'$ and $1e'$ orbitals of BF_3 and the $1a_1$, $2a_1$ and $1e$ orbitals of BF_3^- and BF_3^{2-} to have a constant contribution), the indication is that BF_3^{2-} may be stable, and as a singlet species. The next highest unoccupied orbital for BF_3^{2-} is the 6e orbital that is found to lie very much higher than the $6a_1$ molecular orbital.

From these preliminary calculations for the BF_3 system it is suggested that the BF_3^{2-} ion also may provide an interesting experimental and theoretical investigation. If more sophisticated molecular orbital calculations sustain these preliminary results, it would provide an experimental opportunity to evaluate at least partially the importance of excited states and/or bound states in Z^{2-} ion formation. Thus, if BF_3^{2-} is found to exist experimentally, and other calculations also indicate that the 6e orbital for BF_3^{2-} lies very substantially higher in energy than the $6a_1$ orbital, the importance of excited states attached by this author to the formation and existence of Z^{2-} species may be reduced significantly.

From the above discussions, this author concludes that, in attempting to understand the various Z^{2-} species observed, it is convenient and advantageous to view the Z^{2-} ions generally as existing with relatively long lifetimes in unbound or weakly bound excited states of high multiplicity. A similar viewpoint may be taken with respect to the singly-charged negative ion: the Z^- ion that will yield the Z^{2-} species upon electron capture must have a rather long lifetime and, hence, exist either in an excited state or be stably bound. These views imply, as has been discussed partially above, a generally applicable mechnism for the formation of the Z^{2-} ions.

VI Mechanism of Formation of Doubly-Charged Negative Ions

From the ionization efficiency curve determinations of Stuckey and Kiser, it is known that doubly-charged negative ions are observed only in the energy regions where singly-charged negative ions are produced. This clearly indicated that the singly-charged negative ion is the precursor of the doubly-charged negative ion. In substantiation, the ICR experiments of Stapleton and Bowie have indicated that the Cl^- and Cl^{2-} ions from CCl_4 are coupled.

From the electron beam current variation experiments of Stuckey and Kiser, it is known that two electrons are involved in the overall process of the formation of a doubly-charged negative ion. One of these electrons must be involved in the initial formation of the singly-charged ion by either resonant capture, dissociative resonant capture, or an ion-pair formation process. The second electron then must be gained to form the doubly-charged negative ion in a subsequent step that is energetically realistic.

A single mechanism can now be proposed for the formation of doubly-charged negative ions in the gas phase. This mechanism is consistent with all experimental facts provided to the present, and is as follows.

The first step probably involves the formation of the negative ion in a excited state:

$$M + e \longrightarrow R + Z^{-*} \tag{66a}$$

or $$M + e \longrightarrow R^+ + Z^{-*} + e \tag{66b}$$

or $$Z + e \longrightarrow Z^{-*} \tag{66c}$$

If the species formed in Reactions (66a) through (66c) are produced initially in the ground state, then a second step is required to prepare the Z^- ion in a relatively long-lived excited state:

$$Q + Z^- \longrightarrow Z^{-*} + Q \tag{67a}$$

or $$R^+ + Z^- \longrightarrow Z^{-*} + R^+ \tag{67b}$$

In the final step, electron attachment occurs to form the observed long-lived doubly-charged negative ion:

$$Z^{-*} + e + Q \longrightarrow Z^{2-*} + Q \tag{68}$$

where the electron captured in Reaction (68) generally will be a slow electron. Here, M is a neutral gas molecule and Q is a third body; Q may be the molecule being investigated or a diluent molecule in a plasma.

Reaction (66a) is a dissociative resonant capture. Reaction (66b) is shown as an ion-pair formation process induced by electron impact; Reaction (66c) is electron capture by a molecule or a neutral fragment produced by a previous dissociative process. If the excited state of Z^- lies significantly above the ground state, Reaction (67a) would not be expected to contribute materially to the formation of Z^{-*}. The R^+ involved in Reaction (67b) is either accelerated toward the surface of the plasma or toward the trapping potential plates; this step is an ion-ion reaction in which the necessary excitational energy may be provided the Z^- species. Reaction (68) is a three body process that involves electron attachment to the excited singly-charged negative ion to form the Z^{2-} species observed. In Reaction (68) the Z^{2-} species is indicated as formed in an excited state with a relatively long lifetime.

It is impossible at this stage to specify either the excited states of the Z^{-*} species or of the Z^{2-} ions. In the earlier discussion of electronic states, plausible excited states were indicated that may be involved in these reactions. Quite possibly

146

the Z^{-*} and Z^{2-*} ions are produced in a number of different initial states. What is to be recognized is that excited states of both Z^- and Z^{2-} must be involved to provide for the relatively long lifetimes observed. It is considered reasonable to assume that these excited states are species of high multiplicities. Also it must be recalled that some of the binding of these states may result from the magnetic fields employed[11].

VII Conclusions

Several conclusions may be drawn from the investigations described above. Foremost among these is that doubly-charged negative ions exist. The doubly-charged negative ions may be monatomic, diatomic or polyatomic species; specific cases of each have been reported. These ions exist with lifetimes > 0.01 ms (the flight times of ions in some mass spectrometers) and more of the order of 0.1 ms, the flight times involved in the detection of these species in the omegatron and ion cyclotron resonance spectrometers.

With the exception of the experiments of Ahnell and Koski[4, 5] and Bowie and Stapleton[47], all of the doubly-charged negative atomic ions are reported to have been produced in low temperature plasma ion sources: the omegatron, the ICR spectrometer, the duoplasmatron and the Penning source. These sources involve magnetic fields that may make possible the formation of unbound or weakly bound excited states of high multiplicity (triplet, quartet and possibly quintet) and long lifetimes. This author has suggested in the above discussions a number of plausible excited states in which the doubly-charged negative ions might exist. Massey[196] has indicated that Cl^{2-} might exist as a sextet with the configuration $3p^4(^3P)4s^1 4p^2$ and O^{2-} as a septet with configuration $2p^3(^4S)3s^1 3p^2$. At this time, however, no definitive information is available concerning the states of the doubly-charged negative ions.

The formation of excited species of the Z^- ions is more probable in plasma-type ion sources that involve fairly strong magnetic fields. The atomic species carrying two electronic charges more than the neutral atom are those of the more electronegative elements and the species with the greater electron affinities. The role which nuclear-excited Feshbach resonances may play in the formation of long-lived doubly-charged negative ions is not known. For rather large molecules, separation of charge sites may provide the necessary stabilization and excited states may not play such an important role.

It is certain that our knowledge of doubly-charged negative ions is still very incomplete. The probability is rather large that many additional doubly-charged negative ions remain to be discovered. Some further comments in this vein are given below. The mechanisms suggested in the previous pages are an initial attempt to more fully understand this new class of gaseous negative ions. Additional experiments are now required to characterize more completely and more satisfactorily these species and their mode(s) of formation.

VIII Future Efforts

From the above observations and discussions it is clear that a number of additional studies are warranted. One of these would be the extraction of the relatively slow doubly-charged negative ions from the omegatron into a drift tube so that the lifetimes of the different Z^{2-} species might be determined. Also it would be interesting to learn more about the various modes of decay and the products formed from these Z^{2-} ions.

It is believed likely that many more doubly-charged negative ionic species remain to be detected. One very enticing avenue of investigation will be the search for additional species, e. g., CO_3^{2-}, SO_3^{2-} and general ZF_n^{2-} species, for as more doubly-charged negative ions are detected, their existence and properties must be explainable with any satisfactory theory for the molecular and electronic structures of these entities. Several potential directions for these searches have already been indicated above.

A suggestion that this author made nearly a decade ago[166] still is appropriate, namely, to investigate the negative ion mass spectrum of $Fe(CO)_5$ in search of the species $Fe(CO)_n^{2-}$, where $n \leqslant 4$. Although Compton and Stockdale[67] used a time-of-flight mass spectrometer to investigate the negative ions formed from $Ni(CO)_4$ and $Fe(CO)_5$, even as done earlier by Winters and Kiser[308], the ion source employed by both groups is not suited to the production of large quantities of the doubly-charged negative ions. Since $Fe(CO)_5$ is not particularly stable, the use of a Penning source may be unsatisfactory unless operated at very low arc currents; however, an omegatron or ICR spectrometric study should prove fruitful.

In addition to the omegatron, the ICR, and the Penning ion sources for the production of the doubly-charged negative ions, this author suggests that a "quistor" or quadrupole ion storage source[82, 83, 290] might advantageously be investigated as a possible source for the Z^{2-} species. Parkes[227] has studied negative ion-molecule reactions occurring under swarm conditions, and the quistors have been used to determine rate constants and ion kinetic energies. McGuire and Fortson[203] have used these Penning traps for studying electron-atom collisions at low energy. The quistors can be operated at about 10^4 times lower pressure than used in conventional ion-molecule studies[289]. Such a search with ion trapping for doubly-charged negative species formed in simple systems, such as C_2N_2, NO_2 and SF_6, would, if successful, also provide the opportunity to measure lifetimes and rate constants of the doubly-charged negative ions.

Von Zahn and Tatarczyk[299] have employed a quadrupole beam guide to study metastable lifetimes of ions formed from hydrocarbons. A similar approach to the Z^{2-} ions formed from suitable molecules in a Penning source would permit the determination of the lifetimes of the metastable species observed by Baumann, Heinicke, Kaiser and Bethge[26] in their deflection experiments.

Since the formation of a wide variety of doubly-charged negative monatomic ionic species is possible with a Penning ion source, and since nA beam currents of the Z^{2-} species may be obtained after momentum/charge and kinetic energy analysis with magnetic and electric sectors, photodetachment studies of the Z^{2-} species should be possible to enrich our understanding of these ions and the states in which they are formed and/or exist. Methods such as those noted by Branscomb[48] or the use of kinetic energy analysis of the photodetached electrons, as done by Brehm, Gusinow and Hall[50] and Engelking and Lineberger[103] employing a laser source, should be suitable. Apparently Feldmann has attempted such experiments[153], but without success to the present.

The qualitative investigations of Baumann, Heinicke, Kaiser and Bethge[26], if they can be repeated successfully, should be extended to provide quantitative studies of the pressure effects upon the formation and decomposition of the Z^{2-} species. Thus, the order of Reactions (30) and (32) might be determined. This would provide additional information about the mechanisms of these processes.

Dewar and Rzepa[87] have calculated with the MINDO semiempirical SCF–MO method that $TCNQ^{2-}$ is only about 0.43 eV less stable than $TCNQ^-$, and is more stable than the tetracyanoquinodimethan (TCNQ) molecule. Compton and Cooper[66] make no mention of a doubly-charged negative ion in their report of the negative ion properties of TCNQ. They do, however, comment upon the importance of excited states in complex negative ion systems. An omegatron or ICR search for the $TCNQ^{2-}$ ion would be appropriate, particularly as a test of the utility of this semiempirical approach.

It was mentioned above that the Be^-, BeH^- and Mg^- species were observed by Bethge, Heinicke and Baumann[36]. Weiss[306] has calculated that the states of $2s^1 2p^2\ ^4P\ Be^-$ and $3s^1 3p^2\ ^4P\ Mg^-$ are metastable and have positive electron affinities of 0.24 and 0.32 eV, respectively, relative to the $ns^1 np^1\ ^3P$ state of the neutral atom. Kaufman and Sachs[163] have carried out LCAO–MO–SCF calculations using Gaussian basis functions that indicate BeH^- is bound. These calculations also suggest, based on virtual orbital energies, that the BeH^{2-} ion may be bound. This prediction based on molecular orbital calculations might be verified by searching for and finding the BeH^{2-} ion experimentally. Here the use of a Penning ion source would be more appropriate. An inspection of the figures presented by Bethge, Heinicke and Baumann[25, 36] reveals no indication of BeH^{2-} or the isoelectronic LiH_2^{2-} ion. Possibly CN^{2-} exists in one of the figures[36], but this is uncertain. The fact that these same workers have observed many other doubly-charged negative ions, but did not report having observed BeH^{2-} or LiH_2^{2-}, indicates further molecular orbital calculations are desirable as well.

Certainly detailed theoretical treatments of the stabilities, energetics, and electronic states of the doubly-charged negative ions, both monatomic and polyatomic, are now warranted. It was noted earlier that the EHT calculations employed here are only a zero*th* order approximation; the results obtained, however, now justify an effort to employ more sophisticated approaches. The MINDO and LCAO–MO–SCF calculations mentioned above are a distinct improvement over the EHT calculations. Other calculational techniques should be brought to bear on an improved understanding of doubly-charged negative ions.

Kancerevicius[162] recently calculated with a multiconfiguration approximation that all of the doubly-charged monatomic ions he treated (B^{2-}; C^{2-}, N^{2-}, O^{2-}, Al^{2-}, Si^{2-}, P^{2-} and S^{2-}) are unstable and that they cannot exist in the form of free species. In light of the experimental observations, discussed above, of a number of these doubly-charged monatomic species that are known to possess lifetimes of

149

the order of 0.1 ms in the gas phase, it appears appropriate to encourage theoretical groups to give increased attention to studies of the likely states in which these ions exist. Such theoretical studies are desired both for the simpler monatomic Z^{2-} ions, e. g., O^{2-}, F^{2-}, S^{2-}, Cl^{2-}, P^{2-} and H^{2-}, where the results are likely to be more reliable, and for the more complex polyatomic species.

Berry[30] has pointed out that negative atomic ions pose greater problems for the theorists than positive ions or neutral species. The difficulties for negative molecular ions is greater, and Simons[265] recently has reviewed the theoretical studies in this area. The calculation of properties of the doubly-charged negative ions, both atomic and molecular, surely is still more difficult and challenging, but with the progress made in recent years[265], the prognosis for success in attacking this problem is good.

At the present the roles of negative ions in the Earth's upper atmosphere and in stellar atmospheres are not fully understood. Possibly doubly-charged negative ions also are components of these atmospheres. Therefore, it may be profitable to consider the possible presence of O^{2-}, $NO_2{}^{2-}$, H^{2-} (?), etc., in future studies of the ionosphere and the solar atmosphere.

A decade after the discovery of doubly-charged negative ions, our understanding of this new class of ions is far from complete. Explanations and mechanisms have been formulated that now must be tested and possibly refined. Future efforts of both theoretical and experimental studies shall certainly prove informative.

Acknowledgments. The author thanks Exxon (formerly Esso Research and Engineering Co.), Baton Rouge, Luisiana, for the gift of the ion resonance mass spectrometer[214] which was modified and used in the discovery and earliest investigations of the doubly-charged negative ions. It is a pleasure also to acknowledge stimulating discussions of certain aspects of this work with Professors A. L. Companion and R. E. Knight.

IX References

1. Aberth, W.: J. Chem. Phys. *65*, 4329 (1976)
2. Ahearn, A. J., Hannay, N. B.: J. Chem. Phys. *21*, 119 (1953)
3. Ahlrichs, R.: Chem. Phys. Lett. *34*, 570 (1975)
4. Ahnell, J. E., Koski, W. S.: Nature, (Phys. Sci.) *245*, 30 (1973)
5. Ahnell, J. E., Koski, W. S.: Evidence for Doubly-Charged Negative Fluorine Ions from Electron Bombardment of CF_3Cl, *COO-3283-6*, AEC Contract AT (11–1)–3283, 1973, 5 pp
6. Allred, A. L.: J. Inorg. Nucl. Chem. *17*, 215 (1961)
7. Alpert, D., Buritz, R. S.: J. Appl. Phys. *25*, 202 (1954)
8. Alvarez, L. W., Cornog, R.: Phys. Rev. *56*, 379 (1939)
9. Alvarez, L. W., Cornog, R.: Phys. Rev. *56*, 613 (1939)
10. Anbar, M., Schnitzer, R.: Sciene *191*, 463 (1976)
11. Avron, J., Herbst, I., Simon, B.: Phys. Rev. Lett. *39*, 1068 (1977)
12. Baker, R. F., Tate, J. T.: Phys. Rev. *53*, 683 (1938)
13. Baldeschwieler, J. D.: Science *159*, 263 (1968)
14. Baldeschwieler, J. D., Woodgate, S. S.: Accts. Chem. Res. *4*, 114 (1971)
15. Barber, M., Jennings, K. R., Rhodes, R.: Z. Naturforsch. *22a*, 15 (1967)
16. Bartlett, Jr., J. H.: Nature *125*, 459 (1930)

17. Barton, H. A.: Nature *114*, 826 (1924)
18. Barton, H. A.: Phys. Rev. *25*, 469 (1925)
19. Barton, H. A.: Phys. Rev. *26*, 360 (1925)
20. Bates, D. R.: Proc. Roy. Irish Acad. *51A*, 151 (1947)
21. Bates, D. R., Moiseiwitsch, B. L.: Proc. Phys. Soc. (London) *68A*, 540 (1955)
22. Baughan, E. C.: Trans. Faraday Soc. *55*, 736 (1959)
23. Baughan, E. C.: Trans. Faraday Soc. *55*, 2025 (1959)
24. Baughan, E. C.: Trans. Faraday Soc. *57*, 1863 (1961)
25. Baumann, H., Bethge, K., Heinicke, E.: Nucl. Instrum. Methods *46*, 43 (1967)
26. Baumann, H., Heinicke, E., Kaiser, H. J., Bethge, K.: Nucl. Instrum. Methods *95*, 389 (1971)
27. Bell, R. L.: J. Sci. Instrum. *33*, 269 (1956)
28. Berry, C. E.: Phys. Rev. *85*, 765 (1952)
29. Berry, C. E.: J. Appl. Phys. *25*, 28 (1954)
30. Berry, R. S.: Chem. Rev. *69*, 533 (1969)
31. Berry, R. S., Reimann, C. W.: J. Chem. Phys. *38*, 1540 (1963)
32. Bethe, H.: Z. Physik *57*, 815 (1929)
33. Bethge, K.: Umschau *72*, 24 (1972)
34. Bethge, K.: Proc. Int. Conf. Technol. Electrostatic Accel., (eds.). Aitken, T. W. and Tait, N. R. S., Daresbury Nucl. Phys. Lab., Daresbury, N. Warrington, England, 1973 pp. 371–378
35. Bethge, K., Heinicke, E.: Nucl. Instrum. Methods *30*, 283 (1964)
36. Bethge, K., Heinicke, E., Baumann, H.: Phys. Letters *23*, 542 (1966)
37. Beynon, J. H., Hopkinson, J. A., Lester, G. R.: Int. J. Mass Spectrom. Ion Phys. *2*, 291 (1969)
38. Beynon, J. H., Saunders, R. A., Williams, A. E.: Nature *204*, 67 (1964)
39. Blau, L. M., Novick, R., Weinflash, D.: Phys. Rev. Lett. *24*, 1268 (1970)
40. Bloembergen, N., Purcell, E. M., Pound, R. V.: Phys. Rev. *73*, 679 (1948)
41. Bloom, J. H., Ludington, C. E., Phipps, R. L.: Nuovo Cimento Suppl. *1* (2), 442 (1963)
42. Böhm, H., Günther, K. G.: Z. angew. Physik. *17*, 553 (1964)
43. Bolduc, E., Quéméner, J. J., Marmet, P.: Can. J. Phys. *49*, 3095 (1971)
44. Bouchy-Lorentz, B., Remy, M., Muller, C.: Int. J. Mass Spectrom. Ion Phys. *22*, 147 (1976)
45. Bowie, J. H., Blumenthal, T.: J. Amer. Chem. Soc. *97*, 2959 (1975)
46. Bowie, J. H., Hart, S. G., Blumenthal, T.: Int. J. Mass Spectrom. Ion Phys. *22*, 7 (1976)
47. Bowie, J. H., Stapleton, B. J.: J. Amer. Chem. Soc. *98*, 6480 (1976)
48. Branscomb, L. M.: Photodetachment, in: Atomic and molecular processes, Bates, D. R. (ed.). New York: Academic Press 1962, p. 100
49. Braunbek, W.: Z. Physik, *63*, 20 (1930)
50. Brehm, B., Gusinow, M. A., Hall, J. L.: Phys. Rev. Lett. *19*, 737 (1967)
51. Brodie, I.: Rev. Sci. Instrum. *34*, 1271 (1963)
52. Brubaker, W. M., Perkins, G. D.: Rev. Sci. Instrum. *27*, 720 (1956)
53. Buchel'nikova, I. S.: Zh. Eksp. i Teor. Fiz. *35*, 1119 (1958); [Soviet Phys. – JETP *8*, 783 (1959).]
54. Buchel'nikova, N. S.: Usp. Fiz. Nauk, *65*, 351 (1958); [Adv. Phys. Sci. (Jerusalem) *65*, 243 (1958).]
55. Buchel'nikova, N. S.: Zh. Eksp. i Teor. Fiz. *34*, 519 (1958); [Soviet Phys. – JETP *7*, 358 (1958).]
56. Buchel'nikova, N. S.: Usp. Fiz. Nauk. *65*, 351 (1958). [Translated by Monks, A. L. AEC-tr-3657, Oak Ridge, Tenn. (1959).]
57. Buchel'nikova, N. S.: Fortschr. Physik *8*, 626 (1960)
58. Cantor, S.: J. Chem. Phys. *59*, 5189 (1973)
59. Chait, E. M., Askew, W. B., Matthews, C. B.: Org. Mass Spectrom. *2*, 1135 (1969)
60. Charles, D., Warnecke, Jr., R. J., Marchais, J. C.: Vide *14*, 274 (1959)
61. Christophorou, L. G.: Atomic and molecular radiation physics, New York: Wiley-Interscience 1971, pp. 409–584
62. Christophorou, L. G., Hadjiantoniou, A., Carter, J. G.: J. Chem. Soc., Faraday Trans. *II*, *69*, 1713 (1973)

63. Chupka, W. A., Spence, D., Stevens, C. M.: Search for Long-lived Doubly-Charged Negative Ions, pp. 126–127, Radiological and Environmental Research Division Annual Report, ANL-75-3-PI, July 1973–June 1974, Argonne National Laboratory, Argonne, Ill., 1974

64. Colburn, C. B., Kennedy, A.: J. Amer. Chem. Soc. *80*, 5004 (1958)

65. Compton, R. N.: J. Chem. Phys. *66*, 4478 (1977)

66. Compton, R. N., Cooper, C. D.: J. Chem. Phys. *66*, 4325 (1977)

67. Compton, R. N., Stockdale, J. A. D.: Int. J. Mass Spectrom. Ion Phys. *22*, 47 (1976)

68. Cosby, P. C., Moseley, J. T.: Phys. Rev. Lett. *34*, 1603 (1975)

69. Craggs, J. D., Massey, H. S. W.: The collision of electrons with molecules, in: Handbuch der Physik *37*, (I), 314 (1959)

70. Crossley, R. J. S.: Proc. Phys. Soc. (London) *83*, 375 (1964)

71. Crossley, R. J. S., Coulson, C. A.: Proc. Phys. Soc. (London) *81*, 211 (1963)

72. Cunningham, D. L., Edwards, A. K.: Phys. Rev. *A8*, 2960 (1973)

73. Curran, R. K.: J. Chem. Phys. *38*, 780 (1963)

74. Cusachs, L. C.: J. Chem. Phys. *43*, 5157 (1965)

75. Cusachs, L. C., Corrington, J. H.: Atomic orbitals for semiempirical molecular orbital calculations, in: Sigma molecular orbital theory. by Sinanoglu, O., Wiberg, K. B., New Haven: Yale University Press 1970, pp. 256–272

76. Cusachs, L. C., Reynolds, J. W.: J. Chem. Phys. *43*, S160 (1965)

77. Cusachs, L. C., Reynolds, J. W., Barnard, D.: J. Chem. Phys. *44*, 835 (1966)

78. Daly, N. R., McCormick, A., Powell, R. E.: Rev. Sci. Instrum. *39*, 1163 (1968)

79. Dawson, P. H. (ed.): Quadrupole mass spectrometry and its applications. Amsterdam: Elsevier Scientific Publishing Company 1976, 349 pp

80. Dawton, R. H. V. M: I. E. E. E. Trans. Nucl. Sci. *19*, 231 (1972)

81. de Groot, W.: Philips Techn. Rev. *12*, 65 (1950)

82. Dehmelt, H. G.: Radiofrequency spectroscopy of stored ions. I: Storage. Adv. Atomic Molec. Phys. *3*, 53 (1967), Bates, D. R., Estermann, I. (eds.). New York: Academic Press, Inc.

83. Dehmelt, H. G.: Radiofrequenz spectroscopy of stored ions. II: Spectroscopy, Adv. Atomic Molec. Phys. *5*, 109 (1969), Bate, D. R., Estermann, I. (eds.). New York: Academic Press, Inc.

84. Delgado-Barrio, G., Prat, R. F.: Phys. Rev. *A12*, 2288 (1975)

85. Demkov, Yu. N., Drukarev, G. F.: Zh. Eksp. i Teor. Fiz. *47*, 918 (1964); [Soviet Physics – JETP, *20*, 614 (1965).]

86. Dewar, M. J. S., Harget, A., Haselbach, E.: J. Amer. Chem. Soc. *91*, 7521 (1969)

87. Dewar, M. J. S., Rzepa, H. S.: J. Amer. Chem. Soc. *100*, 784 (1978)

88. Dibeler, V. H., Reese, R. M.: J. Res. Natl. Bur. Stds. *54*, 127 (1955)

89. Dibeler, V. H., Reese, R. M., Mohler, F. L.: J. Res. Natl. Bur. Stds. *57*, 113 (1956)

90. Dillard, J. G.: Chem. Revs. *73*, 589 (1973)

91. Doggett, G., McKendrick, A.: J. Chem. Soc. *(A)*, 825 (1970)

92. Dolder, K. T., Peart, B.: Rep. Progr. Phys. *39*, 693 (1976)

93. Döpel, R.: Ann. Physik. *76*, 1 (1925)

94. Dougherty, R. C.: J. Chem. Phys. *50*, 1896 (1969)

95. Driessler, F., Ahlrichs, R., Staemmler, V., Kutzelnigg, W.: Theoret. Chim. Acta *30*, 315 (1973)

96. Dudley, F. B., Cady, G. H., Crittenden, A. L.: Org. Mass Spectrom. *5*, 953 (1971)

97. Durup, J.: J. Chem. Phys. *65*, 4331 (1976)

98. Dushman, S., Lafferty, J. M.: Scientific foundations of vacuum techniques. New York: John Wiley and Sons, Inc. 1962, p. 315

99. Edlén, B.: J. Chem. Phys. *33*, 98 (1960)

100. Edlén, B.: Phys. Rev. Lett. *28*, 943 (1972)

101. Edwards, A. G.: Brit. J. Appl. Phys. *6*, 44 (1955)

102. Edwards, A. K.: Phys. Rev. *A12*, 1830 (1975)

103. Engelking, P. C., Lineberger, W. C.: J. Chem. Phys. *67*, 1412 (1977)

104. Ennis, Jr. R. M., Schechter, D. E., Thoeming, G., Donnally, B.: I. E. E. E. Trans. Nucl. Sci. *14*, 75 (1967)

105. Estberg, C. N., LaBahn, R. W.: Phys. Rev. Lett. *24*, 1265 (1970)
106. Fano, U., Cooper, J. W.: Phys. Rev. *138*, 400 (1965)
107. Fano, U., Fano, L.: Basic Physics of atoms and molecules. New York: John Wiley and Sons, Inc. 1959, pp. 281–282
108. Farago, P. S.: Free-electron physics. Baltimore: Penguin Books, Inc. 1970, pp. 269
109. Feldman, D., Novick, R.: Observations on the $(1s2s2p)$ $^4P_{5/2}$ metastable state in lithium, in: Atomic collision processes, McDowell, M. R. C. (ed.) Amsterdam: North Holland Publishing Company 1964, pp. 201–210
110. Fox, R. E., Curran, R. K.: J. Chem. Phys. *34*, 1595 (1961)
111. Fraas, R. E.: Ionic Fragmentation Processes of Beta-Diketonate Complexes of the Group III Metals, Doctoral Dissertation, University of Kentucky, Lexington, Kentucky, 1972. 221 pp. (Order no. *73–7,343*, University Microfilms, Ann Arbor, Mich.)
112. Franklin, J. L., Harland, P. W.: Ann. Rev. Phys. Chem. *25*, 485 (1974)
113. Fremlin, J. H.: Nature *212*, 1453 (1966)
114. Geltman, S.: J. Chem. Phys. *25*, 782 (1956)
115. Ginsberg, A. P., Miller, J. M.: J. Inorg. Nucl. Chem. *7*, 351 (1958)
116. Glockler, G.: Phys. Rev. *46*, 111 (1934)
117. Glockler, G.: J. Chem. Phys. *32*, 708 (1960)
118. Glockler, G., Sausville, J. W.: Trans. Electrochem. Soc. *95*, 282 (1949)
119. Goode, G. C., Ferrer-Correia, A. J., Jennings, K. R.: Int. J. Mass Spectrom. Ion. Phys. *5*, 229 (1970)
120. Goode, G. C., O'Malley, R. M., Ferrer-Correia, A. J., Jennings, K. R.: Nature *227*, 1093 (1970)
121. Gross, M. L., Fairweather, R. B., Haddon, W. F., McLafferty, F. W., Major, H. W.: Metastable Ions in Mass Spectra, presented at the 16th Annual Meeting, ASTM Committee E-14 on Mass Spectrometry, Pittsburgh, Pa., 12–17 May 1968
122. Guntz, M.: Compt. rend. *110*, 1337 (1890)
123. Hadjiantoniou, A., Christophorou, L. G., Carter, J. G.: J. Chem. Soc., Faraday Trans. II, *69*, 1691 (1973)
124. Hadjiantoniou, A., Christophorou, L. G., Carter, J. G.: J. Chem. Soc., Faraday Trans. II, *69*, 1704 (1973)
125. Hartmann, H., Lebert, K.-H., Wanczek, K.-P.: Topics Curr. Chem. *43*, 57–115 (1973)
126. Hay, P. J., Dunning, Jr., T. H.: J. Chem. Phys. *67*, 2290 (1977)
127. Hay, P. J., Dunning, Jr., T. H., Goddard, III, W. A.: J. Chem. Phys. *62*, 3912 (1975)
128. Hay, P. J., Goddard, III, W. A.: Chem. Phys. Lett. *14*, 46 (1972)
129. Hebling, A., Lichtman, D.: Fragment Patterns and Appearance Potentials using the Omegatron Mass Spectrometer, presented at the 9th Annual Meeting of ASTM Committee E-14 on Mass Spectrometry, Chicago, Ill., 4–9 June 1961
130. Heinicke, E., Baumann, H.: Nucl. Instrum. Methods *74*, 229 (1969)
131. Heinicke, E., Bethge, K., Baumann, H.: Nucl. Instrum. Methods *58*, 125 (1968)
132. Henis, J. M. S.: Anal. Chem. *41*, (10), 22a (1969)
133. Henis, J. M. S., Mabie, C. A.: J. Chem. Phys. *53*, 2999 (1970)
134. Henrich, L. R.: Astrophys. J. *99*, 59 (1944)
135. Herman, F. H., Skillman, S.: Atomic Structure Calculations, Englewood Cliffs, N. J.: Prentice-Hall, Inc., 1960
136. Herrick, D. R., Stillinger, F. H.: J. Chem. Phys. *62*, 4360 (1975)
137. Herron, J. T., Dibeler, V. H.: J. Chem. Phys. *33*, 1595 (1960)
138. Herzog, R. F.: The monopole: Design and performance, in: Quadrupole mass spectrometry and its applications, Dawson, P. H. (ed.). Amsterdam: Elsevier Scientific Publishing Company, 1976, pp. 153–180
139. Hettich, A.: Z. anorg. Chem. *167*, 67 (1927)
140. Hiby, J. H.: Ann. Physik (5), *34*, 473 (1939)
141. Hickam, W. M., Berg, D.: Negative ion formation and electrical breakdown in some halogenated gases, in: Advances in mass spectrometry, Vol. 1. Waldron, J. D. (ed.). New York: Pergamon Press 1959, pp. 458–472

142. Hill, R. N.: Phys. Rev. Lett, *38*, 643 (1977)
143. Hipple, J. A., Sommer, H., Thomas, H. A.: Phys. Rev. *76*, 1877 (1949)
144. Hipple, J. A., Sommer, H., Thomas, H. A.: Phys. Rev. *78*, 332 (1950)
145. Hobrock, D. L.: Electron Impact Studies of Some Halogen-Containing Hydrocarbons, Doctoral Dissertation, Kansas State University, Manhattan, Kansas, 1965. 142 pp. (Order no *65–12,633* University Microfilms, Ann Arbor, Mich.)
146. Hoffmann, R.: J. Chem. Phys. *39*, 1397 (1963)
147. Hoffmann, R.: J. Chem. Phys. *40*, 2474 (1964)
148. Hoffmann, R.: J. Chem. Phys. *40*, 2480 (1964)
149. Hoffmann, R.: J. Chem. Phys. *40*, 2745 (1964)
150. Holøien, E.: Arch. Math. Naturv. *51*, 81 (1951)
151. Holøien, E., Geltman, S.: Phys. Rev. *153*, 81 (1967)
152. Holøien, E., Midtdal, K.: Proc. Phys. Soc. (London) *68A*, 815 (1955)
153. Hotop, H., Lineberger, W. C.: J. Phys. Chem. Ref. Data *4*, 539 (1975)
154. Huber, B. A., Cosby, Jr., P. C., Peterson, J. R., Moseley, J. T.: J. Chem. Phys. *66*, 4520 (1977)
155. Hull, A. W.: Phys. Rev. *18*, 31 (1921)
156. Iczkowski, R. P., Margrave, J. L.: J. Amer. Chem. Soc. *83*, 3547 (1961)
157. Jaffe, H. H.: J. Chem. Phys. *21*, 258 (1953)
158. Jaffe, H. H., Doak, G. O.: J. Chem. Phys. *21*, 196 (1953)
159. John, J., Robinson, C. P., Aldridge, J. P., Wallace, W. J., Chapman, K. R., Davis, R. H.: I. E. E. E. Trans. Nucl. Sci. *14*, 82 (1967)
160. Johnson, H. R., Rohrlich, F.: Nature *183*, 244 (1959)
161. Johnson, H. R., Rohrlich, F.: J. Chem. Phys. *30*, 1608 (1959)
162. Kancerevicius, A.: Liet. Fiz. Rinkinys. *15*, 215 (1975); [see Chem. Abstr. *83*, 137123a (1975).]
163. Kaufmann, J. J., Sachs, L. M.: J. Chem. Phys. *53*, 446 (1970)
164. Keller, R.: Helv. Phys. Acta *22*, 78 (1949)
165. Khvostenko, V. I., Tolstikov, G. A.: Usp. Khim. *45*, 251 (1976); [Russ. Chem. Revs. *45*, 127 (1976).]
166. Kiser, R. W.: Mass spectrometry of organometallic compounds, in: Characterization of organometallic compounds, Part 1, Tsutsui, M. (ed.). New York: Interscience Publishers 1969, p. 202
167. Kiser, R. W.: Introduction to mass spectrometry and its applications. J.: Prentice-Hall, Inc., Englewood Cliffs, N. 1965, 356 pp.
168. Kiser, R. W., Sullivan, R. E., Lupin, M. S.: Anal. Chem. *41*, 1958 (1969)
169. Kleinberg, J., Argersinger, Jr., W. J., Griswold, E.: Inorganic chemistry, Boston: D. C. Heath and Co. 1960, p. 603
170. Klopfer, A.: Das Omegatron als Partialdruckmesser, presented at the 1st International Congress on Vacuum Techniques, Belgium 10–13 June 1958. [see Vacuum *9*, 78 (1959).]
171. Klopfer, A., Schmidt, W.: Vacuum *10*, 363 (1960)
172. Klopfer, A., Schmidt, W.: Philips Techn. Rev. *22*, 195 (1960)
173. Knight, R. E., Scherr, C. W.: Rev. Mod. Phys. *35*, 431 (1963)
174. Koopmans, T. A.: Physica *1*, 104 (1933)
175. Ladd, M. F. C., Lee, W. H.: Acta Crystallog. *13*, 959 (1960)
176. Laughlin, C., Stewart, A. L.: Proc. Phys. Soc. (London) *88A*, 893 (1966)
177. Lawrence, E. O., Livingston, M. S.: Phys. Rev. *40*, 19 (1932)
178. Lehman, T. A., Bursey, M. M.: Ion cyclotron resonance spectrometry, New York: Wiley-Interscience 1976, 230 pp.
179. Levy-Leblond, J. M.: J. Phys. *B, 4*, L23 (1971)
180. Lichtman, D.: J. Appl. Phys. *31*, 1213 (1960)
181. Lifshitz, C., Wu, R. L. C., Tiernan, T. O., Terwilliger, D. T.: Chem. Phys. *68*, 247 (1978)
182. Lin, C. D.: Phys. Rev. *A12*, 493 (1975)
183. Llewellyn, P. M.: Ion Cyclotron Resonance Mass Spectrometer, 13th Annual Conference on Mass Spectrometry and Allied Topics, ASTM Committee E-14 on Mass Spectrometry, St. Louis, Mo., 16–21 May 1965

184. Lofthus, A.: Mol. Phys. *5*, 105 (1962)
185. Lorrain, P.: Can. J. Res. *25A*, 338 (1947)
186. Loughran, E. D., Mader, C.: J. Chem. Phys. *32*, 1578 (1960)
187. Lowe, J. P.: J. Amer. Chem. Soc. *99*, 5557 (1977)
188. Lucchese, R. R., Schaefer, III, H. F.: J. Chem. Phys. *67*, 848 (1977)
189. Mader, D. L., Novick, R.: Phys. Rev. Lett. *29*, 199 (1972)
190. Mader, D. L., Novick, R.: Phys. Rev. Lett. *32*, 185 (1974)
191. Manson, S. T.: Phys. Rev. *A3*, 147 (1971)
192. Marmet, P., Bolduc, E., Quéméner, J. J.: J. Chem. Phys. *56*, 3463 (1972)
193. Marriott, J., Thorburn, R., Craggs, J. D.: Proc. Phys. Soc. (London) *B67*, 437 (1954)
194. Masica, B.: Nuovo Cimento, Suppl. *1* (2), 435 (1963)
195. Massey, H. S. W.: Negative Ions, 2nd edit. London: Cambridge at the University Press 1950, p. 22
196. Massey, H.: Negative Ions, 3rd edit. London: Cambridge University Press, 1976, 741 pp
197. Massey, H. S. W.: Endeavour *35*, (125), 58 (1976)
198. Matese, J. J., Rountree, S. P., Henry, R. J. W.: Phys. Rev. *A8*, 2965 (1973)
199. Maxwell, L. R.: Rev. Sci. Instrum. *2*, 129 (1931)
200. Mazunov, V. A., Khvostenko, V. I.: Khim. Vys. Energ. *10*, 279 (1976); [High Energy Chem. *10*, 253 (1976); UDC-539. 196, VINITI-200-76]
201. McDaniel, E. W.: Collision phenomena in ionized gases. New York: John Wiley and Sons, Inc. 1964 pp. 368 and 376
202. McDaniel, E. W., Mason, E. A.: The mobility and diffusion of ions in gases. New York: John Wiley and Sons, 1973, 372 pp
203. McGuire, M. D., Fortson, E. N.: Phys. Rev. Lett. *33*, 737 (1974)
204. McMahon, T. B., Beauchamp, J. L.: Rev. Sci. Instrum. *42*, 1632 (1971)
205. McNarry, L. R.: Can. J. Phys. *36*, 1710 (1958)
206. Melton, C. E.: Negative ion Mass spectra, in: Mass spectrometry of organic ions, McLafferty, F. W. (ed.). New York: Academic Press 1963, pp. 163–205
207. Melton, C. E.: Principles of mass spectrometry and negative ions. New York: Marcel Dekker, Inc. 1970, 313 pp
208. Messmer, R. P., Salahub, D. R.: J. Chem. Phys. *65*, 779 (1976)
209. Middleton, R.: Nucl. Instrum. Methods *122*, 35 (1974)
210. Moiseiwitsch, B. L.: Proc. Phys. Soc. (London) *67A*, 25 (1954)
211. Moiseiwitsch, B. L.: Electron affinities of atoms and molecules. Adv. Atomic Molec. Phys. *1*, 61 (1965), Bates, D. R., Estermann, I. (eds.). New York: Academic Press, Inc.
212. Moiseiwitsch, B. L.: Negative ions, in: Atomic processes and applications, Burke, P. G, Moiseiwitsch, B. L. (eds.). Amsterdam: North-Holland Publishing Company 1976, pp. 291–319
213. Moore, C. E.: Atomic energy levels, in three volumes, National Bureau of Standards Circular *467*, Washington, D. C., 1949, 1952 and 1958. 303, 229 and 245 pp
214. Morgan, W. A., Jernakoff, G., Lanneau, K. P.: Ind. Eng. Chem. *46*, 1404 (1954)
215. Moser, C. M., Nesbet, R. K.: Phys. Rev. *A11*, 1157 (1975)
216. Muller, C., Lorentz, B., Remy, M.: Int. J. Mass Spectrom. Ion Phys. *18*, 33 (1975)
217. Mulliken, R. S.: J. Chim. Phys. *46*, 497 (1949)
218. Mulliken, R. S.: J. Chim. Phys. *46*, 675 (1949)
219. Mulliken, R. S., Rieke, C. A., Orloff, H., Orloff, D.: J. Chem. Phys. *17*, 1248 (1949)
220. Nasser, E.: Fundamentals of gaseous ionization and plasma electronics. New York: Wiley-Interscience, 1971, p. 128
221. Nazaroff, G. V.: J. Chem. Phys. *59*, 4009 (1973)
222. Nesbet, R. K.: Phys. Rev. *A14*, 1326 (1976)
223. Nicholas, D. J., Trowbridge, C. W., Allen, W. D.: Phys. Rev. *167*, 38 (1968)
224. Novick, R., Weinflash, D.: Precision Measurement of the Fine Structure and Lifetimes of the $(1s2s2p)$ ^4P_J States of He$^-$ and Li*, in: Precision Measurement and Fundamental Constants, National Bureau of Standards Special Publication *343*, Langenberg, D. N., Taylor, B. N. (eds.). *53* pp, August 1971. Washington, D. C. pp. 403–410

225. Page, F. M., Goode, G. C.: Negative ions and the magnetron, New York: Wiley-Interscience, 1969, 156 pp
226. Parkes, G. D.: Mellor's modern inorganic chemistry. New York: John Wiley and Sons, Inc., 1961, p. 668
227. Parkes, D. A.: Vacuum 24, 561 (1974)
228. Paul, W., Steinwedel, H.: Z. Naturforsch. 8a, 448 (1953)
229. Paul, W., Reinhard, H. P., von Zahn, U.: Z. Physik 152, 143 (1958)
230. Pauling, L.: Nature of the chemical bond, 2rd edit. Ithaca, N. Y.: Cornell University Press, 1948, p. 421
231. Pauling, L.: Nature of the chemical bond, 3rd edit. Ithaca, N. Y.: Cornell University Press, 1960, 644 pp
232. Peart, B., Dolder, K. T.: J. Phys. B, 6, 1497 (1973)
233. Pekeris, C. L.: Phys. Rev. 126, 1470 (1962)
234. Penning, F. M.: Physica 4, 71 (1937)
235. Penning, F. M.: Philips Techn. Rev. 2, 201 (1937)
236. Penning, F. M., Moubis, J. H. A.: Physica 4, 1190 (1937)
237. Penning, F. M., Nienhuis, K.: Philips Techn. Rev. 11, 116 (1949)
238. Peper, J.: Philips Techn. Rev. 19, 218 (1957)
239. Pietenpol, J. L.: Phys. Rev. Lett. 7, 64 (1961)
240. Pirani, M., Yarwood, J.: Principles of vacuum engineering, 2nd impression, New York: Reinhold Publishing Corporation, 1963, 578 pp
241. Prat, R. F.: Phys. Rev. A6, 1735 (1972)
242. Pritchard, H. O.: Chem. Rev. 52, 529 (1953)
243. Purser, K. H.: I. E. E. E. Trans. Nucl. Sci. 20, 136 (1973)
244. Quéméner, J. J., Paquet, C., Marmet, P.: Phys. Rev. A4, 494 (1971)
245. Redhead, P. A.: Trans. Roy. Soc. Can. 47, 134 (1953)
246. Reese, R. M., Dibeler, V. H., Mohler, F. L.: J. Res. Nat. Bur. Stds. 57, 367 (1956)
247. Rehkopf, C. H.: Nuovo Cimento, Suppl. 1 (2), 749 (1963)
248. Reich, G., Flecken, F.: Vacuum 10, 35 (1960)
249. Reichert, C., Fraas, R. E., Kiser, R. W.: Int. J. Mass Spectrom. Ion Phys. 5, 457 (1970)
250. Riviere, A. C., Sweetman, D. R.: Phys. Rev. Lett. 5, 560 (1960)
251. Rose, F. A., Tollefsrud, P. B., Richards, H. T.: I. E. E. E. Trans. Nucl. Sci. 14, 78 (1967)
252. Schaefer, III, H. F., Heil, T. G.: J. Chem. Phys. 54, 2573 (1971)
253. Scherr, C. W., Knight, R. E.: Rev. Mod. Phys. 35, 436 (1963)
254. Scherr, C. W., Silverman, J. N., Matsen, F. A.: Phys. Rev. 127, 830 (1962)
255. Schnitzer, R., Anbar, M.: J. Chem. Phys. 64, 2466 (1976)
256. Schnitzer, R., Anbar, M.: J. Chem. Phys. 65, 4332 (1976)
257. Schoen, R. I.: Can. J. Chem. 47, 1879 (1969)
258. Schuchhardt, G.: Vacuum 10, 373 (1960)
259. Schulz, G. J.: Rev. Modern Phys. 45, 423 (1973)
260. Shadoff, L. A.: Anal. Chem. 39, 1902 (1967)
261. Sheehan, W. F.: J. Amer. Chem. Soc. 100, 1348 (1978)
262. Shih, S., Buenker, R. J., Peyerimhoff, S. D.: Chem. Phys. Lett. 28, 463 (1974)
263. Sidgwick, N. V.: The chemical elements and their compounds, Vol. 1. Oxford: Clarendon Press 1950, p. 121
264. Siegbahn, K., Nordling, C., Johannson, G., Hedman, J., Hedén, P. F., Hamrin, K., Gelius, U., Bergmark, T., Werme, L. O., Manne, R., Baer, Y.: ESCA applied to free molecules. Amsterdam: North-Holland Publishing Company, 1969, pp. 94–98
265. Simons, J.: Theoretical Studies of Negative Molecular Ions, Ann. Rev. Phys. Chem. 28, 15 (1977)
266. Simpson, F. R., Browning, R., Gilbody, H. B.: J. Phys. B, 4, 106 (1971)
267. Slater, J. C.: Phys. Rev. 36, 57 (1930)
268. Smirnov, B. M.: Dokl. Akad. SSSR 161, 92 (1965); [Soviet Physics – Doklady 10, 218 (1965).]
269. Smirnov, B. M.: Teplofiz. Vys. Temp. 3, 775 (1965); [High Temp. 3, 716 (1965).]

270. Smirnov, B. M., Chibisov, M. I.: Zh. Eksp. i Teor. Fiz. *49*, 841 (1965); [Soviet Physics – JETP, *22*, 585 (1966).]
271. So, S. P.: J. Chem. Phys. *67*, 2929 (1977)
272. Sommer, H., Thomas, H. A.: Phys. Rev. *78*, 806 (1950)
273. Sommer, H., Thomas, H. A., Hipple, J. A.: Phys. Rev. *82*, 697 (1951)
274. Spence, D., Schulz, G. J.: J. Chem. Phys. *58*, 1800 (1973)
275. Stapleton, B. J., Bowie, J. H.: Austral. J. Chem. *30*, 417 (1977)
276. Stark, D. S.: Vacuum *9*, 288 (1959)
277. Steckelmacher, W., Buckingham, J. D.: Nuovo Cimento, Suppl. *1 (2)*, 418 (1963)
278. Steiner, B.: Photodetachment: cross sections and electron affinities, in: Case studies in atomic collision physics II. by McDaniel, E. W., McDowell, M. R. C. (eds.). Amsterdam: North-Holland Publishing Company 1972, pp. 483–545
279. Steinmetz, H., Hettich, A.: Z. anorg. Chem. *167*, 75 (1927)
280. Stuckey, W. K.: The Formation and Study of Gaseous Doubly-Charged Negative Ions in an Omegatron Mass Spectrometer, Doctoral Dissertation, Kansas State University, Manhattan, Kansas, 1966. 111 pp. (Order no. 66–10,740, University Microfilms, Ann Arbor, Mich.)
281. Stuckey, W. K., Kiser, R. W.: Nature *211*, 963 (1966)
282. Svec, H. J., Flesch, G. D.: Int. J. Mass Spectrom. Ion Phys. *1*, 41 (1968)
283. Sweetman, D. R.: Proc. Phys. Soc. (London) *76*, 998 (1960)
284. Tang, S. Y., Rothe, E. W., Reck, G. P.: Int. J. Mass Spectrom. Ion Phys. *14*, 79 (1974)
285. Taylor, H. S., Thomas, L. D.: Phys. Rev. Lett. *28*, 1091 (1972)
286. Thomas, L. D.: J. Phys. *B*, *7*, L97 (1974)
287. Thomson, J. J.: Rays of positive electricity and their application to chemical analyses. London: Longmans, Green and Co., 1913, p. 18 and pp. 39–46
288. Thorne, P. C. L., Roberts, E. R.: Ephraim's inorganic chemistry, 5th edit. London: Gurney and Jackson, 1948, p. 262
289. Todd, J. F. J.: Applications in atomic and molecular physics, in: Quadrupole mass spectrometry and its applications. Dawson, P. H. (ed.). Amsterdam: Elsevier Scientific Publishing Company, 1976, pp. 241–271
290. Todd, J. F. J., Lawson, G., Bonner, R. F.: Quadrupole ion Traps, in: Quadrupole mass spectrometry and its applications. Dawson, P. H. (ed.). Amsterdam: Elsevier Scientific Publishing Company, 1976, pp. 181–224
291. Truell, R.: Proc. Inst. Radio Engrs. *36*, 1249 (1948)
292. Truell, R.: Proc. Inst. Radio Engrs. *37*, 1144 (1949)
293. Tüxen, O.: Z. Physik *103*, 463 (1936)
294. van der Waal, J.: Nuovo Cimento Suppl. *1 (2)*, 760 (1963)
295. van der Waal, J., Francken, J. C.: Philips Techn. Rev. *23*, 122 (1961)
296. Vestal, M. L.: J. Chem. Phys. *65*, 4331 (1976)
297. von Ardenne, M., Steinfelder, R., Tümmler, R.: Elektronenanlagerungs-Massenspektrographie organischer Substanzen. Berlin, Heidelberg, New York: Springer, 1971, 403 pp
298. von Niessen, W., Cederbaum, L. S., Diercksen, G. H. F., Hohlneicher, G.: Chem. Phys. *11*, 399 (1975)
299. von Zahn, U., Tatarczyk, H.: Phys. Letters *12*, 190 (1964)
300. Wagener, J. S., Marth, P. T.: J. Appl. Phys. *28*, 1027 (1957)
301. Wagman, D. D., Evans, W. H., Halow, I., Parker, V. B., Bailey, S. M. Schumm, R. H., Churney, K. L.: Selected Values of Chemical Thermodynamic Properties, National Bureau of Standards Technical Notes 270–1 through 270–6, 1965, 1966, 1968, 1969 and 1971. U. S. Government Printing Office, Washington, D. C.
302. Walton, D. S., Peart, B., Dolder, K. T.: J. Phys. *B*, *3*, L148 (1970)
303. Walton, D. S., Peart, B., Dolder, K. T.: J. Phys. *B*, *4*, 1343 (1971)
304. Warnecke, Jr., R. J.: Vacuum *10*, 49 (1960)
305. Warnecke, Jr., R. J., Marchais, J. C.: Vide *16*, 114 (1961)
306. Weiss, A. W.: Phys. Rev. *166*, 70 (1968)
307. Windham, P. M., Joseph, P. J., Weinman, J. A.: Phys. Rev. *109*, 1193 (1958)
308. Winters, R. E., Kiser, R. W.: J. Chem. Phys. *44*, 1964 (1966)

R. W. Kiser

309. Wobschall, D.: Rev. Sci. Instrum. *36*, 466 (1965)
310. Wobschall, D., Graham, Jr., J. R., Malone, D. P.: Phys. Rev. *131*, 1565 (1963)
311. Wobschall, D., Graham, Jr., J. R., Malone, D. P.: J. Chem. Phys. *42*, 3955 (1965)
312. Wöhler, L.: Z. anorg. Chem. *78*, 239 (1912)
313. Wolfsberg, M., Helmholz, L.: J. Chem. Phys. *20*, 837 (1952)
314. Wu, T.-Y.: Phil. Mag. *22*, 837 (1936)
315. Wu, T.-Y., Shen, S. T.: Chinese J. Phys. *5*, 150 (1944); [see Chem. Abstr. *40*, 2736 (1945).]
316. Younkin, J. M., Smith, L. J., Compton, R. N.: Theoret. Chim. Acta *41*, 157 (1976)
317. Zdanuk, E. J., Bierig, R., Rubin, L. G., Wolsky, S. I.: Vacuum *10*, 382 (1960)
318. Zollweg, R. J.: J. Chem. Phys. *50*, 4251 (1969)

Received June 8, 1978

Chromatography in Inorganic Trace Analysis

G. Schwedt

Gesamthochschule Siegen, Naturwissenschaften II, Analytische Chemie, 5900 Siegen 21, Germany

Table of Contents

Introduction

Compared with spectroscopic and electrochemical methods, chromatographic methods have been little used so far for routine quantitative inorganic trace analysis. However, gas chromatography of volatile metal chelates has given rise to increasing application possibilities for inorganic trace analysis. Development of instruments and technology for detection as well as the manufacture of separating material have also made column fluid chromatography suitable for detection of element traces. Higher selectivities and higher sensitivities than with other methods of element trace analysis can be achieved today for some elements using chromatographic methods. Further advantages of chromatographic methods are in the

separation of elements which interfere with detection,
separation of the matrix,
separation of excess reagent,
separation of mixtures of element traces (multielement analysis),
reduction of the work involved in sample preparation in consequence of separations and
linking of separation with sensitive and selective methods of detection.

Analysis data in the extreme trace range are frequently subject to great uncertainty with regard to their accuracy when they have only been determined according to one method. Chromatographic methods are thus suitable as reference methods.

1 Gas-Chromatography

The separations of many volatile substances for inorganic analysis has been described in the literature[1-4]. However, only a small proportion of these publications deal with the possibilities and applications of trace analysis. The reasons for this lie in the difficulty of sensitive detection of inorganic compounds, their instability and the low volatility of many of them, and in the numerous interfering effects in the chromatographic column.

In spite of these difficulties gas-chromatography offers some interesting possibilities for inorganic trace analysis.

In many cases the necessity of chemical reactions of cations and anions to volatile derivatives for gas-chromatiography leads to a convenient separation from diverse matrices for the subsequent determination. With the aid of gas-chromatography, analyses can be carried out over a wide range of concentrations usually by simple changing the detection system while maintaining the same chromatographic conditions. The tendency of substances with the same ligand or organic group such as metal chelates or organometallic compounds to separate offers possibilities for multielement analysis. Gas-chromatographic techniques can be utilized as reference methods, especially for the ultra-trace analysis of ppb- and lower levels.

1.1 Gas-Chromatographic Separations with Quantitative Analyses

The use of gas-chromatography in inorganic analysis requires that the compounds be thermally stable and that they do not react with the different parts of the chromatographic system itself (injection unit, column, detector). Moreover they should be formed quantitative by a suitable reaction in the analytical range. Gaseous elements, other inorganic gases, hydrogen compounds and halogen compounds of elements are suitable for determinations in the ppm- and ppb-range (Table 1.1). Metal halogenides can also be determined sensitively with different detectors (Table 1.1), but difficulties in the trace field are found because of the sensitivity of these substances to hydrolysis.

Table 1.1. Gas chromatographic separations with quantitative analyses

Element	Compound (or ligand)	Detector	Detectable amount [g]	Ref.
I. Inorganice compounds				
As	$AsCl_3$	ECD	10^{-10}	51)
Cyanide	$CN \cdot Cl$	ECD	2.5×10^{-10}	113)
Sn	$SnCl_4$	TCD	4×10^{-6}	14)
Ti	$TiCl_4$	FPD	$10^8 - 10^{-9}$	56)
II. Fluorinated β-diketonates				
Be	(tfa)	ECD	4×10^{-13}	78)
Be	(tfa)	TCD	4×10^{-8}	97)
Cr	(hfa)	ECD	10^{-11}	76)
Ni	(Monothiotrifluoracetylacetone)	ECD	5×10^{-11}	13)
Th, U	(Heptafluoro-β-diketones)	FID	10^{-9}	39)
V	(tfa)	FID/ECD	$10^{-7}/10^{-10}$	53)
Cr, Al, Rh	(hfa, tfa)	ECD	$10^{-10} - 10^{-12}$	81)
Cu, Ni, Pd, Pt	(β-Diketones, β-thioketones, bidentate and tetradentate β-ketoamines)	ECD	10^{-9}	16)
Metals	(Heptafluorodimethyloctene-dione)	FID	ca. 10^{-9}	95)
Pb	Perfluoralcoylpivalmethanes	MS	10^{-14}	17)
III. Mixed ligand complexes				
Ni, Co	(tfa + DMF)	FID	4×10^{-7}	52)
Mn, Fe, Co, Ni	(hfa + DBSO)	FID	$6 - 11, 5 \times 10^{-8}$	72)
Zn	$Zn(hfa)_2$ 2 DBSO	FID	4×10^{-8}	73)
U, Th	(hfa + DBSO)	TCD	$4 \times 10^{-7}, 6 \times 10^{-7}$	53)
Cerium group lanthanides	(Decafluoro-3,5-heptanedione + DBSO)	FID	10^{-7}	28)
Yttrium group lanthanides	(Decafluoro-3,5-heptanedione + DBSO)	ECD	2×10^{-7}	27)

Table 1.1. (continued)

Element	Compound (or ligand)	Detector	Detectable amount [g]	Ref.

IV. Organometallic compounds

Element	Compound (or ligand)	Detector	Detectable amount [g]	Ref.
As	Triphenylarsine	FID	10^{-11}	99)
Hg	Aryl-, alkylmercuryhylides	ECD	10^{-11}	117)
Hg	Phenylmercurychloride	ECD	4.5×10^{-11}	62)
Hg	Phenylmercurychloride	ECD	4.5×10^{-10}	54)
Cl	Phenylmercurychloride	FID	4×10^{-10}	18)
Tl	Tl(I)cyclopentadienyl	TCD	10^{-6}	20)

V. Miscellaneous compounds

Element	Compound (or ligand)	Detector	Detectable amount [g]	Ref.
F	Triethylfluorsilane	TCD/FID	$2 \times 10^{-7}/6 \times 10^{-10}$	21)
I	Iodoacetone	ECD	2×10^{-10}	48)
Se	5-Nitropiazselenol	ECD	4×10^{-10}	68)
Metals	Simple and fluorinated diethyl-dithiocarbamates	FID/ECD	ng-levels	105, 106)

VI. Indirect determinations

Element	Compound (or ligand)	Detector	Detectable amount [g]	Ref.
Ru	As hexachlorobicycloheptene-2,3-γ-lactone	ECD	2×10^{-10}	12)
Ni	As polychlorinated alcohol (from xanthate)	ECD	10^{-11}	96)

Abbreviations: tfa: trifluoracetylacetone; hfa: hexafluoroacetylacetone; DMF: dimethylform-amide; DBSO: di-n-butylsulfoxide — TCD: thermal conductivity detector; for other detector abbreviations see Table 1.2.

The group of metal chelates with β-diketones and similar ligands can be extracted from weakly acid aqueous solutions under set pH conditions with an organic solvent, in which the chelate-forming substance is soluted. Excess reagent can be re-extracted in the aqueous phase with diluted sodium hydroxide; the metal chelate remains in the organic solvent. Polar metal chelates require packed columns with deactivated supporting materials because of their temperature sensitivity.

By use of different detectors trace amounts of metals ranging from 10^{-8} to 10^{-14} g can be determined (Tables 1.1/II, 1.2). Polyfluorinated β-diketones as ligands give more highly volatile derivatives and above all allow the use of the electron capture detector (ECD) which has low detection limits for fluorinated compounds. Nevertheless decompositions can also appear[39] in this class of substances. In addition thioketones[16] and amino derivatives[16, 111, 112] are used as chelating agents for quantitative gas-chromatography.

Table 1.2. Gas-chromatography: special detectors

Detector	Compound	Detectable amount [g]	Ref.
FID/ECD (comparison)	$Cr(tfa)_3$	$6 \times 10^{-8}/$ 2.1×10^{-10}	9)
ECD/coulometric d. /FPD/ MS	Boron hydrides	(ppb-Levels)	100)
Radiometric detector	SO_2 as $SO_2{}^{18}F_2$	10^{-18}	11)
Radiometric detector	^{3}H-Labelled β-diketonates (Cr, Al)	ca. 10^{-12}	26)
FPD	P	10^{-12}	5)
FPD	S, SO_2	5×10^{-9}	35)
FPD	Selenium compounds	2×10^{-10}	37)
FPD	$Cr(hfa)_3$	10^{-6}	50)
FPD	Metal halides, metal chelates (Ti)	ca. 10^{-9}	56)
MS (integrated current method)	Pb perfluoralcoylpivalmethanes	10^{-14}	17)
MS	Metal polyfluoro-β-diketonates	10^{-14}	59)
MED	Cr trifluoro-2,4-pentenedione	9×10^{-13}	19)
MED	Metal acetylacetonates (e. g. Sc)	ca. 10^{-11}	33)
MED	Metal acetylacetonates (Al, Be, Cr)	$10^{-7}, 10^{-11}, 10^{-9}$	58)
MED	Organic mercury compounds	10^{-12}	103)
AAS	Dimethylselenide, dimethyldiselenide	10^{-10}	31)
AAS (flame and flameless atomization)	$SbCl_3$ in $GaCl_3$	ca. $50 \times$ or 3×10^{-12}	82)
AAS	$Cr(tfa)_3$	10^{-9}	115)

Abbreviations: FID: flame ionization detector; ECD: electron capture detector; FPD: flame photometric detector; MS: mass spectrometer; MED: microwave emission detector; AAS: atomic absorption spectrometer.

Another group of compounds which have recently proved interesting for gas-chromatography are ternary complexes from fluorinated β-diketones and organic solvent molecules, such as dimethylformamide or di-n-butylsulfoxide (Table 1.1/III). For these compounds synergetic effects through primary separation of metals by liquid-liquid extraction can be profitable[73]; in some cases separations can be achieved more conveniently with gas-chromatography[72]. For quantitative analysis of zinc using gas-chromatography was made possible for the first time through its ternary complex[73]. Nevertheless Pb and Cd compounds of this type (Table 1.1/III) are composed. In general in the gas-chromatographic column these ternary complexes show a higher stability than binary complexes. Interactions between polar metal chelates and the packed chromatographic column, which produce elution curves with tailing, can be successfully reduced by the presence of a carrier gas containing ligand vapour[101].

Additional substance groups suitable for gas chromatography are organometallic compounds (Table 1.1/IV), which can yield from inorganic compounds especially for Hg and As by means of different alkylating and arylating reagents such as Grignard compounds[98, 99, 102], tetraphenylborate[18, 62, 117], Li pentafluorobenzenesulphinate[66], pentacyanomethylcobaltate(III) and Na 4.4-dimethyl-4-silanepentene-1-sulphonate[55].

In addition anions such as fluoride and iodide can be determined using gas-chromatography, when they appear as part of organic compounds[21, 48]; oxygen-containing anions such as phosphate, arsenite and selenite can be changed into volatile derivatives with silylating reagents[29, 60, 63]. Piazselenoles have been used in the determination of selenium, in the picogram and nanogram range the contents of the chromatographic columns are also analyzed to determine losses[64]. The procedure is similar to that described for organic Hg compounds[62].

Recently sulfur-containing metal chelates and nonfluorinated diethyldithiocarbamates of relatively low volatility have been successfully used to determine some metals for gas-chromatographic multielement analysis[34, 75, 105, 106]. For As carbamate only the careful deactivation of the column with silylating agents allows quantitative gas-chromatographic analysis[34].

For non-volatile metal chelates the practibility of indirect determination overcomes if the substances are to be decomposed (pyrrolized) in definited products. Instances for this approach are given, using Ru and Ni. The metal chelates are first separated by thin-layer chromatography[12, 96] (Table 1.1/IV).

The greatest problem in the application of gas-chromatographic methods to elemental trace analysis is the interaction between the compound used for analysis and the column packing: conditioning of the column and the application of chelating agents in the gas phase are ways of diminishing these effects. For additional details about packing materials see Ref.[1, 3].

1.2 Special Detectors

In addition to the thermal conductivity detector (TCD), flame photometric (FPD) and atomic absorption spectrometric (AAS) detectors, which offer high sensitivity and the advantage of selectivity, are also suitable for use with inorganic substances (Table 1.2). By-products from chemical reactions of the analytical process will not be detected.

The flame ionization detector (FID), which, for fluorinated compounds, has the advantage of a greater linearity range in spite of its lower sensitivity in comparison with the electron capture detector (ECD), has been extensively applied to organic metal chelates in particular. Comparisons of the different detectors were carried out (Table 1.2)[9, 100]. For special analyses such as the determination of SO_2 by reaction to SO_2F_2 with the radioactive ^{18}F-isotope or by utilization of ^3H-labelled β-diketonates, radiometric measurements for detection are employed[11, 26].

Mass spectrometric detection, which is widely used in organic trace analysis, allows sensitive and selective determinations, but high cost and the unresolved interfacing problems limit its application (e.g.[17, 59]).

As well as the atomic spectroscopic methods of flame photometry and atomic absorption spectroscopy microwave emission spectroscopic detection (MED) is being used more and more. MED combines high sensitivity in the picogram range with high selectivity for elemental analysis. It is as suitable for inorganic and organic compounds as the two other detectors. The sensitivities for polyfluorinated β-diketonates are situated in the same order of magnitude and are lower in comparison with the ECD[17, 59].

1.3 Applications

The number of publications dealing with chromatography as applied to inorganic analysis now amounts to more than a thousand[1, 4]. The main point of interest in these papers, however, is substance separations. The use of gas-chromatography for elemental trace analysis is only of interest if, in addition to quantitative results for test solutions — as described and summarized in Section 1.1 and 1.2 — information about its application to different materials is also submitted. Such information is presented in Tables 1.3.1 and 1.3.2.

It is difficult to compare the detection limit data because most of the publications do not make clear how the values were obtained. For trace analytical procedures in general the detection limit is calculated from a quantity measured at the detection limit, which is in turn determined from the mean value of the blank value plus three-times its standard deviation.

A chromatogram is interpreted not only by the peak area but also by the signal (peak) height. The detection limit of an analytical procedure is given by concentration levels as ppm or ppb (10^{-6} or 10^{-8} parts per gram material).

A quantitative recovery of the determining element is rarely achieved as result of necessary clean-up procedures and chemical derivatizations. The range and re-

Table 1.3. Recoveries and standard deviations for some gas chromatographic applications in inorganic trace analysis

Element (ion)	Analyzed compound	Material	Recovery (mean value)	[Conc.] — s_{rel} [%]	Ref.
As	Arsenic diethyldithiocar-bamate	Water, urine	82.6	[1 ppm] — 9.3	34)
As	Triphenylarsine	Liver	100	[0.4 ppm] — 10	98)
Be	Be(tfa)$_2$	Whole blood	92	[1.14 ppm] — 7.1	107)
Cr	Cr(tfa)$_3$	Plasma	90.5	[5.16 ppb] — 7.5	19)
Hg	Phenylmercurychloride	Water	70.5	[1 ppm] — 9.6	66)
		Urine	81.4	[1 ppm] — 12.7	
		Serum	51.0	[1 ppm] — 18.4	
Hg	Phenylmercurychloride	Serum	80.4	[3.6 ppm] — 9.7	117)
		Liver	93.4	[3.6 ppm] — 8.3	
Ni	Ni monothiotrifluoro-acetylacetonate	Tea	94	[13 ppm] — 10.8	13)
Se	5-Nitropiazselenol	Pure elemental arsenic	100	[3.76 ppm] — 4.8	8)
Fluoride	Triethylfluorsilane	Liver	83	[2.5 ppm] — 20	83)
Nitrate	Triethoxynitrobenzene	Water	81	[0.2 ppm] — 3.7	110)

producibility of recovery of some elements or ions, for which gas-chromatographic determination has been applied, are compiled in Table 1.3. The achieved relative standard deviations in the ppm-range are comparable with those of other trace analytical methods.

In some publications the application of gas-chromatography as a reference method has been compared with other analytical methods. The gas-chromatographic determination of aluminum and chromium in uranium gave yields which agreed with those of spectrophotometry in the ppm range[43]. Comparable results were also obtained for beryllium yields below one ppm in standard probes with a fluorimetric method (with Morin as reagent)[36]. In both instances for gas-chromatographic determination the trifluoroacetylacetone chelates were used. For yields of 0,02 ppm Be in NBS standard reference Orchard Leaves it was not possible to determine beryllium because of the presence of very large interfering peaks close to the beryllium peak[38]. Therefore only emission spectrometric or fluorimetric methods could be used. The results of the determination of chromium in biological materials were compared with those from determination by atomic absorption spectrometry after extraction as Cr(VI) into methylisobutylketone[87]. The slope of the regression line was 0.901 with a correlation coefficient of 0.963 for ppb-levels.

For mercury determination in sediments or fish tissues, samples were oxidized with boiling nitric acid. After reagent addition the methyl mercury derivate was extracted by benzene. The results of gas-chromatographic determination were higher than those obtained by wet-oxidation and atomic absorption methods[55].

1.3.1 Inorganic Materials (see Table 1.3.1)

For determination of beryllium in inorganic materials, the samples are fused with sodium carbonate and then dissolved in hydrochloric acid. Beryllium trifluoroacetylacetonate is formed and extracted by adding a benzene solution of the ligand after adjusting the pH and adding EDTA-solution for complexing of other metals[36, 38]. An aliquot part of the $Be(tfa)_2$-containing benzene layer is submitted to gas-chromatographic analysis. In ferrous alloys chromium also is determined as $Cr(tfa)_3$. In the presence of catalytic amounts of nitric acid, the reaction with trifluoroacetylacetone is stimulated with microwave radiofrequency energy without organic solvent[80].

The determination of fluoride in water and inorganic phosphates is described after its conversion to triethylfluorsilane by reaction with triethylchlorsilane in a hydrochloric acid medium[21]. The silane derivative is extracted into tetrachloroethylene.

A sensitive gas-chromatographic determination of selenium is possible as 5-chlor- or 5-nitro-piazselenol. These compounds are extractable into benzene or toluene for direct injection into a gas-chromatograph with ECD or TCD[7, 8, 64, 90]. After treatment of steel samples with diluted mixtures of equal volumes of hydrochloric and nitric acids in the presence of perchloric acid, the selenium is completely converted into the quadrivalent state. Large amounts of Fe(III) are masked with phosphoric acid[7]. In pure sulfuric acid ultramicro amounts of selenium are converted into seleneous acid with a bromine-bromide redox buffer solution[90]. For high-purity

Table 1.3.1. Gas-chromatography — applications for inorganic materials

Element	Material	Compound	Detector	Detection limit	Ref.
As	Oxides, sulphides, ores and alloys	$AsCl_3$	TCD	ppm-Range	51)
Be	Lunar samples, terrestrial meteorites	$Be(tfa)_2$	ECD	40 ppb	36, 94)
Be	NBS standards	$Be(tfa)_2$	ECD	ppb-Levels	38)
Cr	Ferrous alloys	$Cr(tfa)_3$	ECD	20 ppb	80)
F	Water, phosphates	Triethylfluorosilane	FID	ppm to ppb-levels	21)
Se	Steel	5-Chloropiaz-selenol	TCD	ca. 0.1%	7)
Se	Pure elemental arsenic, arsenic(III)-oxide	5-Nitropiazselenol	ECD	< 3 ppm	8)
Se	High-purity copper	5-Nitropiazselenol	ECD	1 ppb	64)
Se	Pure sulfuric acid	5-Nitropiazselenol	ECD	ppm-Levels	90)
Si	Iron and steel	$SiCl_4$	Gas density balance d.	50 ppm	92)
Sn	Zircaloy	$SnCl_4$	TCD	ppm-Levels	14)
Al, Cr	Uranium	tfa-Compounds	ECD	0.1 ppm	43)
Cl, Si	$SiBr_4$	Cl, $SiCl_4$	MS	0.2 ppm	6)
C, S	Helium	CO_2, CO, COS, H_2S	ECD (helium detector	ppm-Range	24)
Cu, Al	Zinc	tfa-Compounds	MED	20 ppb	85)
NO_3^-, NO_2^-	Aqueous test solutions	Nitrobenzene	ECD	0.12 ppm	44)

copper the sample is heated in a quartz tube at $1,100-1,150°$ in an oxygen-stream. SeO_2 evaporates and is collected quantitatively by condensation of the carrier gas in a micro-trap cooled with liquid nitrogen. After evaporation of the oxygen, SeO_2 is made to react in the trap with 4-nitro-1,2-diaminobenzene and is then extracted with toluene[64]. For determination of silicium, the samples in a graphite boat were heated at 650° in chlorine stream. Volatile chlorides were trapped in a packed U-tube cooled at −5°. After pumping with nitrogen the trap was heated by a water bath to 90−100°, and the volatile $SiCl_4$ and other chlorides were swept into the gas-chromatography column with a recovery of 97% (92). Also zirconium-tin alloy (Zircaloys) samples were submitted to a reaction with gaseous chlorine in a similar way. An apparatus for this method described[14]. After removing the zirconium chloride and an excess of chlorine, the tin tetrachloride is determined by gas chromatography. Arsenic was determined in the form of arsenic chloride after the quantitative chlorination of different inorganic materials with carbon tetrachloride in sealed glass tubes. The volatile products were introduced into the gas chromatographic system after crushing the tube in a special device[51].

A solution of uranyl nitrate was used for the extraction of aluminum and chromium from acetate-buffered aqueous media with trifluoracetylacetone in benzene.

After shaking for 30 minutes the organic phase was washed with diluted ammonia to re-extract the reagent[43]. For analysis of copper and aluminum in zinc with the microwave plasma emission detector (MED) the metal was dissolved in aqua regia, the solution was evaporated to dryness, and the residue was dissolved in dilute hydrochloric acid. At pH 4.5 the complexes were extracted with trifluoracetylacetone in chloroform. Enrichment by evaporation of the organic layer is possible[85]. The anions nitrite and nitrate are treated in aqueous solutions with benzene and H_2SO_4-H_2O (3:1) to convert them (nitrite ions after oxidation with aqueous $KMnO_4$) into nitrobenzene[44]. The recovery of nitrate is 90 (+8)% within a range of 0.12–62 ppm; those of nitrite 93 (+12)% (range 0.09–37 ppm).

Inorganic impurities in samples of helium were analyzed by gas-chromatography with a laboratory-constructed detector, which was described[24].

The impurities in silicium tetrabromide were identified[6]. The packing irreversibly adsorbs $SiBr_4$; methane and chlorine were not separated, but they could be determined simultaneously by using FID for CH_4 and an ECD for chlorine.

1.3.2 Organic and Environmental Materials (see Table 1.3.2)

In elemental trace analysis gas-chromatography especially has been established as reliable for those elements which can be determined only with lower sensitivities

Table 1.3.2. Gas-chromatography – applications for organic and environmental materials

Element	Material	Compound	Detector	Detection limit	Ref.
Al	Water	Al(tfa)$_3$	ECD	< 1 ppm	61)
Al	Rat liver	Al(tfa)$_3$	ECD	ppb-Levels	65)
As	Water, urine	As Diethyldithiocarbamate	ECD	1 ppm	34)
As	Biological materials	TMS-arsenate	FID	0.1 ppm	60)
As	Liver	Triphenylarsine	FID	0.25 ppm	98)
Be	Biological materials	Be(tfa)$_2$	ECD	10 ppb	57)
Be	Urine	Be(tfa)$_2$	ECD	1 ppb	40)
Be	Biological samples	Be(tfa)$_2$	ECD	ppm-Levels	41)
Be	Biological materials, air	Be(tfa)$_2$	ECD	ca. 1 ppb	71)
Be	Whole blood, tissue homogenates	Be(tfa)$_2$	ECD	20 ppb	107)
Be	Biological fluids	Be(tfa)$_2$	ECD	2 ppb	108)
Br	Blood	1,2-Dibromocyclohexane	FID	(0.1 μg/ml)	10)
Cl	Water	Phenylmercury(II)-chloride	FID	8 ppb	18)
Cr	Human plasma	Cr(tfa)$_3$	MED	5 ppb	19)
Cr	Biological tissues (liver)	Cr(tfa)$_3$	ECD	20 ppb	23)
Cr	Blood/plasma	Cr(tfa)$_3$	ECD	5 ppb	47)

Table 1.3.2. (continued)

Element	Material	Compound	Detector	Detection limit	Ref.
Cr	Biological materials	$Cr(tfa)_3$	ECD	ppb-Levels	87)
Cr	Serum	$Cr(tfa)_3$	ECD	ca. 10 ppb	88)
Ge	Coal	$GeCl_4$	Catharo-meter	3.3 ppm	89)
F	Water	Triethylfluorsilane	FID	30 ppb	22)
F	Urine, serum, saliva, bone	Trimethylfluorsilane	FID	10 ppb	42)
F	Biological materials	Triethylfluorsilane	FID	1 ppm	83)
Hg	Water	Methylmercury compounds	ECD	2.5 ppb	55)
Hg	Water, urine, serum	Arylmercury compounds	ECD	20 ppb	66)
Hg	Biological materials	Methylmercurychloride	ECD	1 ppb	30)
I	Milk	Iodoacetone	ECD	ca. 3 ppb	46)
P	Water, mud, biological samples	Phosphorous	FPD	ppb-Levels	5)
P	Water	TMS-phosphate	FID, FPD	0.1 ppm	63)
Ni	Tea, fats	Ni monothiofluoro-acetylacetonate	ECD	ppb-Levels	13)
S	Air	SO_2	FPD	5 ppb	25)
Se	Biological materials	5-Nitropiazselenol	ECD	5 ppb	74)
Se	Milk, milk products, albumin	5-Nitropiazselenol	ECD	5 ppb	91)
Se	Environmental samples (water/solid samples)	Piazselenol	MED	0.1/15 ppb	104)
Se	Urine	Piazselenol	ECD	Ca. 7 ppb	116)
As-Sb	Environmental samples (water/solid samples)	Triphenyl compounds	MED	50–125/ 30–75 ppb	102)
Al-Cr	Water	tfa-Compounds	ECD	4–0.5 ppb	45)
Cu-Ni-Zn	Marine bottom sediments	Diethyldithiocar-bamates	FID	Ca. 1 ppm	75)
CN^-, SCN^-	Water	$CN \cdot Br$	ECD	10 ppb	69, 70)
CN^-	Water	$CN \cdot Br$	ECD	ppm-Levels	79)
HCN	Combustion effluents	HCN	FID	1 ppm	67)
CN^-	Biological specimen	$CN \cdot Cl$	ECD	25 ppb	113)
NO_3^-, NO_2^-	Saliva, blood, water, air	Nitrobenzene	ECD	0.1 ppm	110)
H_2O	Organic solvents	H_2O	TCD	2 ppm	86)
CO	Air	CO	TCD	< 1 ppm	109)

and higher detection limits by other methods: These are, in particular, aluminum, chromium, beryllium and selenium as well as the anions chloride, iodide and nitrate (Table 1.3.2).

For the determination of aluminum sea water was shaken for one our with tri-fluoracetylacetone in toluene. Unreacted ligand was removed with dilute ammonia solution. Samples with high dissolved organic content, which caused the ECD to function abnormally, were oxidized by a standard ultraviolet irradiation technique before chelation and extraction steps[61]. Rat liver was digested with hot H_2SO_4-HNO_3, and the digest was adjusted with aqueous sodium hydroxide to pH 4.5 for the extraction of aluminum with trifluoracetylacetone[65].

Samples of water or urine were treated with sodium iodide in hydrochloric acid to generate arsenic(III) and then they were treated with diethyldithiocarbamate, followed by isolation into toluene[34]. An interesting feature of the chromatographic behavior of the complex is the necessity for extensive silanization of the packing and column wall to permit satisfactory results. For gas-chromatographic determination of arsenic as trimethylsilylarsenate, organic matter was destroyed by acid oxidation in a Teflon bomb. After extraction by diethyldithiocarbamate, oxidation and silylation by methyl-trimethylsilyl-heptafluorobutyramide arsenic can be determined in the ppm-range[60]. From arsenic diethyldithiocarbamate triphenylarsane was also quantitatively formed by a reaction with magnesium diphenyl. Triphenyl-arsane is particulary suitable for gas-chromatography with the FID. Dried biological material is incinerated in a Schöniger flask[98].

For the determination of beryllium, different organic materials were destroyed by low temperatur ashing or pressure decomposition with nitric acid/hydrofluoric acid in a Teflon tube. Interfacing elements were masked or pre-extracted; at pH 9 the beryllium trifluoroacetyl-acetonate was formed and extracted into benzene. The concentration of $Be(tfa)_2$ solutions is possible[57]. From biological fluids such as urine beryllium is directly extracted as $Fe(tfa)_2$ [40, 71, 107, 108]. Direct dissolution-chelation of beryllium with concentrated trifluoracetylacetone suffered from interference by a tfa reaction product. A second method employed the dissolution of BeO in dog blood and rat liver homogenate by a hot 75% sodium hydroxide procedure[41]. The dissolved beryllium oxide was then chelated by alow temperature reaction without an interfering product. Comparative radiochemical and GC analyses were made.

In a protein-free filtrate of blood bromide was oxidized to bromine by potassium permanganate in acid solution and extracted into cyclohexane-containing cyclohexene to give 1,2-dibromocyclohexane[10, 114].

Chloride in aqueous solutions is readily converted to phenylmercury(II)chloride by treatment with phenylmercury(II)nitrate at pH 1.5 and the covalently bound chloride can be quantitatively extracted into chloroform[18]. Before reproducible peaks could be achieved, it was necessary to make four injections, because there was some absorption on the stationary phase[64].

For determination of chromium two different forms for derivatization are described. Following wet digestion of tissue the solution was incubated at pH 5.8—6.1 with trifluoracetylacetone in benzene with shaking for one hour at $70°$[23, 87, 88]. In the gas-chromatogram two peaks of both *trans-* and *cis-*isomers of $Cr(tfa)_3$ were observed[87]. In a direct reaction procedure 0.05 ml samples of blood or plasma

$Cr(tfa)_3$ were formed at $175°$ (within 30 minutes): it was not necessary to digest or ash the sample[47]. Further applications for the determination of chromium are described in Ref.[83].

Treatment of coal with gaseous hydrogen chloride at $600–700°$ effects complete extraction of germanium tetrachloride, whereas heating the coal at $300°$ liberates volatile organic compounds that interfere with gas chromatographic analysis[89]. Preliminary treatment with concentrated acids (e. g. sulfuric) at $150°$ also allows a high extraction, if a concentrator consisting of a U-tube packed with activated carbon is used.

Traces of fluoride can be separated from aqueous solutions by extraction with triethylchlorsilane in m-xylene[22]. The conditions were optimized and co-precipitation (for enrichment) reactions were tested[22]. Adsorption on hydroxyl apatite was found most suitable. From acidified solutions of biological materials fluoride was selectively extracted with a solution of trimethylchlorsilane in benzene[42]. Organic material or blood was destroyed by heating with hydrochloric acid and hydrogen peroxide in a closed distilling apparatus. The resulting distillates were extracted by triethylchlorsilane in tetrachloroethylene[83].

Mercury(II) solutions were mixed with 2.5 M sulfuric acid, 15% aqueous sodium nitrite and 1% aqueous sodium 4,4-dimethyl-4-silapentene-1-sulfonate for the conversion of mercury into a methyl mercury derivative by heating the tube in a boiling water bath for ten minutes[55]. After cooling the extraction with benzene follows. For application to sediment or fish tissues the samples were oxidized with boiling nitric acid. To prevent the occurence of interfering peaks, the benzene solution was extracted with sodium thiosulfate solution; potassium iodide was added, and the mercury compound was then extracted in toluene[55]. As arylating agent lithium pentafluorobenzenesulfinate gives good results for the determination of inorganic mercury in water, urine and serum[66].

Inorganic mercury was also isolated as methyl mercury upon reaction with tetramethyl tin[30]. The initial extracts were subjected to thiosulfate clean-up, and mercury was isolated as the bromide derivative. The method yielded good agreement between gas-chromatographic and atomic absorption spectrometric data. Twenty-four samples were analyzed daily on a routine basis. Organomercurials could also be determined and the differences from inorganic mercury could be detected by these gas-chromatographic methods[30].

Various forms of iodine are convertible into iodoacetone, which was extracted with hexane[46]. For the detection of iodide in milk, proteins were precipitated with diluted sulfuric acid, and the filtrate was treated with acetone and potassium iodate. For total iodine a digestion with potassium carbonate was carried out at $600°$[46]. Elemental phosphorus is extracted into benzene or isooctane from water, aqueous filtered extractions of mud or homogenized tissue-filtered extracts[5].

Trace phosphate determination is effected by extraction with an organic solution of a quaternary ammonium salt, silylation of the extracted phosphate directly in the organic layer and gas-chromatographic measurement with FID or FPD[63].

After wet or dry ashing of tea samples such as mineralizing of fats by dry ashing procedure, trace amounts of nickel can extracted at pH 4.5–5.0 with a solution

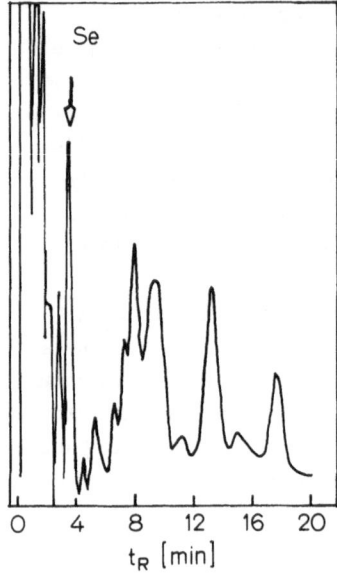

Fig. 1.1. Gas chromatogram of the determination of selenium as 5-nitropiazselenol from dried skim milk, detector: ECD (after Shimoishi 1976, Ref.[91])

of monothiotrifluoroacetylacetone in n-hexane[13]. The extract was washed with sodium hydroxide solution to destroy residual chelating agent.

After nitric acid-magnesium nitrate digestion selenium can be derivated to 5-nitropiazselenole and determined after extraction into toluene[74]. The method is suitable for the determination in orchard leaves, bovine liver, human placenta, hair, blood and urine. Similar digestion was used for the gas-chromatographic determination of selenium in milk, milk products and albumin (Fig. 1.1)[91]. In environmentally-based samples after wet ashing selenium was also determined as 5-nitropiazselenole by a gas-chromatograph with a microwave emission spectrometric detector (MED)[104]. Matrix effects, column deterioration and selenium loss were examined. The determination of the physiological levels of selenium in human blood, urine and river water was employed after its reaction with 2,3-diamino-naphthalene at pH 2 and extraction of the piazselenol into hexane[116]. Digestion was effected with boiling nitric acid.

With gas-chromatography it is possible to determine the presence of more than one element. The determination of arsenic and antimony as triphenyl compounds is based on their co-crystallization with thionalid and reaction of the precipitate with phenylmagnesiumbromide (Grignard)[102]. Following the decomposition of reagent excess the compounds were extracted into diethylether. Atomic emission is measured by a MED. Solid samples of coal, fly ash, orchard leaves and bovine liver (0,2–1 g) were digested by a nitric acid-perchloric acid mixture wet ashing procedure[102, 98, 99].

The simultaneous analysis of aluminum and chromium from water was effected by heating samples with trifluoroacetylacetone in benzene at 55–60° for 20–25 minutes and shaking the organic layer with dilute sodium hydroxide[45]. High sensitivities were reached for both elements (Fig. 1.2).

The metal diethyldithiocarbamates[105, 106] were used to achieve the simultaneous determination of copper, nickel and zinc in marine bottom sediments[75]. Powdered

Fig. 1.2. Gas-chromatographic determination of Al and Cr as trifluoroacetyl-acetonates, chromatogram from 6.12×10^{-13} g Al and 9.12×10^{-13} g Cr, detector: ECD (after Gosink 1975, Ref.[45])

and dried samples were treated in a micro-autoclave with mixtures of nitric and hydrogen fluoride acids at 150° for thirty minutes. The digestion is evaporated to remove silicon compounds. After precipitation of iron and manganese from the filtrate of the ammonia solution the elements were extracted as carbamates (Fig. 3).

Other anions were also sensitively and selectively determined by gas-chromatography. Cyanide and thiocyanate were determined from cyanogen bromide after treatment of ortho-phosphoric acid solutions with bromine water[69]. The excess of bromine was removed by adding aqueous phenol. For complex cyanides the first step was the breakdown by irradiation with ultraviolet light[70]. An aliquot part of the aqueous sample was injected directly into the gas-chromatograph. Using chloramine-T (sodium para-toluene sulfonechloramide) for the conversion of cyanide to cyanogen chloride, it was possible to determine this anion in the biological specimen[113]. Cya-

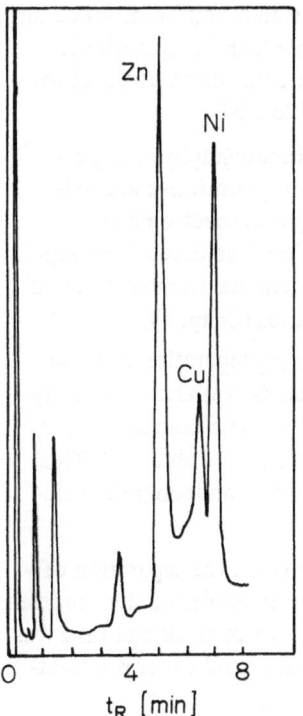

Fig. 1.3. Determination of Zn, Cu and Ni as diethyldithiocarbamates by gas-chromatography from a marine bottom sediment sample (concentrations in the sample: 17.1 ppm Zn – 1.6 ppm Cu – 4.9 ppm Ni), detector: FID (after Radecki et al. 1978, Ref.[75])

173

nogen chloride was extraceted into hexane. Cyanide was separated from blood, urine and aqueous solutions by a technique using microdiffusion cells containing 0.1 N sodium hydroxide in the central reservoir[113]. For gas-chromatographic determination of hydrogen cyanide combustion effluents were carried directly from a flow reactor into a gas-sampling valve[67].

In Table 1.3.2 a method for the analysis of nitrates and nitrites for a wide variety of samples is given[110]. Aqueous nitrate ions are converted to nitrobenzene by reaction with benzene in the presence of concentrated sulfuric acid as catalyst. The special methods for the different materials are described. To stabilize samples against bacterial action, which can reduce nitrate concentrations, phenylmercuric acetate is added after collection[79, 110]. Other inorganic compounds in the trace levels for gas-chromatographic determination are, for example, water and carbon monoxide, which were determined in organic solvents or air in the ppb-range[86, 109].

2 Liquid Column Chromatography

Ion-exchange chromatography has been extensively extended, particulary in the area of inorganic analysis. It is used for different problems in trace analysis. In many cases ion-exchange chromatography is not used to separate a substance mixture into single compounds, but to separate out one ion from many others, which would interfere with its quantitative analysis (Table 2.1.1). The determination of the separated ion then follows after the ion-exchange step in the eluate fraction by different procedures. Over the last few years methods have been described, with which the quantitative analysis of the separated ions is carried out directly by the connection of the chromatographic column with a continuous-flow detector (Table 2.3).

A second extended function of the liquid column chromatography is to pre-separate trace amounts of several substances for subsequent quantitative analysis by a selective determination method such as atomic absorption spectrometry (Table 2.3). Here often only a particular degree of separation is achieved. Non separated elements are to be determined with high-selective methods. In most cases an enrichment is combined with these chromatographic methods (Chap. 4).

In contrast to the first and second examples — concerning separation of trace amounts at the stationary phase — pre-separation of the matrix was also sucessfully undertaken. The ions to be determined were not adsorbed in the chromatographic column, but were placed in the effluent after the adsorption step (Table 2.2). This was advantageous, because losses of trace amounts as a result of an incomplete elution process from the stationary phase were avoided.

However, chromatographic methods are only effective when the separation of a mixture into its single components is possible. Only a few examples of this are given for conventional liquid column chromatography in which a lot of work and time was expended as a result of numerous elution steps, pre-extractions and other pre-treatments (Table 2.4).

Table 2.1.1. Liquid column chromatographic pre-separations of single compounds

Element	Separation method	Material	Detection	Detection limit	Ref.
Ag	Ion-exchange	Water	AAS	0.1 ppb	124)
B	Ion-exchange	Water, nutrient media	Photometry	μg-Levels	153)
Ba	Ion-exchange	Water	AAS	10 ppm	168)
Be	Ion-exchange	Geological, industrial materials	AAS	ppb-Range	143)
Bi	Reversed-phase of thiourea complex (tri-butylphosphate)	Cerussite	AAS	ppm-Range	175)
Ce	Reversed-phase: bis-(2-ethylhexyl)phos-phate	Sea water	Fluorimetry: + Ti^{3+}	1 ng/l	167)
In	Reversed-phase: tri-butylphosphate	Gallium	Fluorimetry; + Rhodamin G	18 ppm	135)
Sn	Asbestos, cellulose: Sn-catechol-complex from catechol violet	Foodstuff (organotin stabilizer)	Photometry	80 ppm	119)
Sn	Cellulose	Tin-copper mixtures	Volumetry	μg-Range	127)
Sn	Ion-exchange	Rocks, soils, red clay, ferromanganese materials	Photometry: + phenylfluorone	0.1 ppm	171)
Tl	Ion-exchange	Silicate rocks, marine sediments, sea water	Fluorimetry, neutron activation	0.1 ppm	151)
U	Reversed-phase: 2-thenoyltrifluoro-acetone	Fission and activation products	Neutron activation	0.5 ng	185)

Recently important advances have made by combining high efficient liquid column chromatography with sensitive detection methods. These techniques are differentiated into two groups:

1. forced-flow liquid chromatography,
2. high-pressure (or high-performance) liquid chromatography (= HPLC).

Forced-flow chromatography can be regarded as a preliminary stage of the HPLC. Ion-exchange separations and spectrophotometric analysis have been coupled by continuous monitoring of column effluent. Moreover, it is differentiated from the classic column chromatography by the utilization of smaller packing particles. In comparison with HPLC in the stationary phase, materials with diameters of more than 50 μm are used which require only lower pressures of 5–10 bar for the promotion of the mobile phase. Because it uses particles with diameters as small as 5 μm and consequently higher pressure, the chromatography is named HPLC[161]. This classification, however, seems arbitrary, and both conditions can be considered together as part of a fast and high-efficiency liquid column chromatography with two distinguishing characteristics in comparison to the classic LC:

1. smaller particle sizes for the stationary phase and therefore higher separation efficiency
2. continuous detection of the effluent

The increase in efficiency has opened possibilities for multielement analyses, as already described for gas-chromatography (Chap. 1).

In Table 2.3 only those publications are classified as forced-flow chromatography which have been explicitly designated as such by their authors. However, earlier publications are also included under this heading. The different chromatographic methods which employ in columns, such as ion-exchange, adsorption, distribution and reversed-phase chromatography allow the separation of several different substance classes. For this reason liquid column chromatography significantly extends the application of chromatography as compared with gas-chromatography which is limited to volatile compounds. In the following sections only those publications pertaining to trace analysis with applications for organic and inorganic materials will be described.

2.1 Liquid Column Chromatographic Separations and Pre-separations (Tables 2.1.1 and 2.1.2)

For determination of silver in water, pre-separation with quantitative elution by an anion-exchange column was used[124]. After chelation with ammonium pyrrolidine dithiocarbamate and extraction of the chelate with MIBK, measurement of silver followed by AAS.

Boron was removed from distilled water by using a column of Dowex-1 (OH^--form). For the photometric determination as a boron-curcumin complex, interfering ions were removed on small columns of Dowex-1(formiate) and Dowex-50 (Na^+)[153]. For the evaporation and ashing step, calcium hydroxide was used (recovery of one μg: ca. 80%). By complexation of bivalent ions with 1,2-diamino-cyclohexane-N,N,N′,N′-tetra-acetic acid, water samples at pH 5.5 passed through a column of Dowex 50 W-X 8 (NH_4^+) for the determination of barium[168]. Elution followed with 4 M nitric acid. Particular references for the separation of barium from sodium, calcium and sulfate are given.

After extraction from the matrix elements by chloroform of its acetylacetonate, beryllium was separated from co-extracted aluminum by means of a column of the strongly acidic cation-exchanger Dowex 50[143]. Adsorption is affected by tetrahydro-furan-chlorform-methanol-hydrochloric acid medium, elution of aluminum with oxalic acid, and beryllium with 6 M hydrochloric acid. Bismuth was selectively adsorbed in the form of a thiourea complex on a column of a polymer with tri-butylphosphate in the presence of the remaining Fe, Sb, Cu[175].

Quadrivalent cerium was precipitated with $Fe(OH)_3$; after dissolving in hydrocloric acid interfering ions were removed by extraction with MIBK[167]. After evaporation with $HClO_4$ the residue was dissolved in water and passed through a column of bis(2-ethylhexyl)-phosphate on PVC, from which cerium was eluted with 0.3 M $HClO_4$.

Table 2.1.2. Liquid column-chromatography for separations of trace components from matrices

Elements	Separation method	Material	Detection	Detection limit	Ref.
Trace el.	Ion-exchange: step-wise elution	Natural waters	AAS, fluorimetry	ppb-Levels	142)
Metals	Ion-exchange: step-wise elution	Manganese modules	AAS	ppm-Range	144)
Metals	Cation-exchange: stepwise elut.	Silicates	AAS	ppm-Range	152)
Cr, Cu, Pb, Fe	Ion-exchange	Synthetic mixtures	Volumetry	10^{-6} N	158)
Cr, V	Chelating resin for VO_3^-, CrO_4^-	–	Photometry	μg-Amounts	149)
Mo, W	Ion-exchange	Sea water	Photometry	ppb-Levels (Mo) < ppb-Levels (W)	139)
Zr, Hf	Reversed-phase: acetophenone	–	Photometry	–	125)
Metals	Ion-exchange	Titandioxide crystals	Ge(Li)spectro-methods	$10^{-10} - 10^{-5}$ %	156)
Metals	Ion-exchange	Silicate rocks	Different methods	ppm-Range	177, 178)
Cu, Ni, Zn, Cd	Chelex 100	Silicate rock samples	AAS	ppm-Range	180)
Zn, Pb, Bi	Ion-exchange	Heat resistance alloys, ferroalloys	Photometry	2 ppm	146)
Pt-group	Ion-exchange	Meteorites	Photometry	ppm-ppb Levels	163)
Cu, Zn	Ion-exchange	Plant leaves	Neutron activation	ppm-Range	172)
Pb, Cd	Ion-exchange	Urine, blood	AAS	< 1 ppm	184)
Pb, U, Th	Ion-exchange	Tantalar-niobates	Complexometry, photometry	ppm-Range	173)
V, Al	Ion-exchange	Titanium	Neutron activation	0.15 ppm	155)
Lu, Yb, Tb	Ion-exchange	Rocks	Neutron activation	ppm-ppb Levels	122)
Cu, Cd, Pb	Cation-exchange as Complexes	Pure bismuth	Stripping volta-metry	10 ppm	169)

Indium was sorbed and separated from germanium from 0.8 M HBr ad the material "Ftoroplast-4", treated with tri-butylphosphate, and eluted with water for fluorimetric determination with Rhodamine G[135].

Organotin-stabiliser can be extracted from foodstuffs by various chlorinated solvents and hydrocarbons. Interfering substances, especially inorganic tin, were pre-separated by paper chromatography[119]. After fusion and formation of the tin-catechol complex, chromatography was used for separation of the tin-complex from excess reagent.

By elution with butanol saturated with N HCl for tin, and with butanol-concentrated. HCL (1:1) for copper, these elements were separable by a cellulose column[127]. For the pre-separation and selective separation of thallium an anion-exchange scheme has been developed[151]. The method showed a coefficient of variation of 1.2% at the 6 ppm level.

With 2-theonyltrifluoroacetone-impregnated Porapak for the determination of U^{239}, Np (from ^{239}U) was separated from fission and activation products and eluted with 10 M HNO_3 with a chemical yield of 96.8 ± 1.6% compared with 97.8 ± 1.5% by a solvent-extraction method[185].

Toxic trace elements were isolated from water samples by extraction with di-ethyldithiocarbamate (Table 2.1.2)[142]. Following this pre-concentration step the metal ions were adsorbed on a cation-exchange resin using a mixture of tetrahydro-furan-methylglycol-6 M HCl as sorption solution. The succesive elution was treated with 6 M HCl, 1 M HCl and 2 M HNO_3 for fractional separation. In another application hexane-isopropanol-HCl mixture was used as the adsorption medium[144]. An analytical scheme which provides quantitative results, is described for ion-exchange separation of fifteen major, minor and trace elements in silicates[152]. For concentration and separation of copper, chromium, lead and iron an ion-exchanger in phosphate or OH^--form was used in various combinations[158].

Using a Dowex A-1 resin, which has a polystyrene matrix with iminodiacetic chelating groups, peroxy compounds of Cr and V were separable by adsorption, while Nb, Ta and W were not adsorbed[149].

Molybdenum and tungsten were quantitatively separated from sea water by the addition of H_2O_2 and NH_4SCN in 0.1 M HCl (for Mo) at a cation-exchange resin[139]. Elution followed with 0.5 M NaOH/0.5 M NaCl.

By using acetophenone as a stationary phase on granular polyethylene, zirconium and hafnium were adsorbed and separately eluted[125].

Combination of anion-exchange and extraction by tri-butylphosphate allowed a rapid and accurate determination of up to 21 metallic impurities in titan dioxide crystals[156]. Anion- and cation-exchange resins were collected in series for separating of the major and some minor elements in silicate materials[177].

By stepwise changing of eluting agents with only a single cation-exchange resin the quantiative separation and accurate determination of ten elements in silicate rocks was possible[178].

Silicates were decomposed with HF and aqua regia in a sealed Teflon vessel[180]. With a chelating resin heavy metals were adsorbed from solutions with malonic acid and eluted with 2 M HNO_3.

Trace amounts of Zn, Pb and Bi (2–200 ppm) in Fe and Ni based alloys were separated by anion-exchange chromatography[146].

Further examples for separations and pre-separations of elements are given in Refs.[163, 172, 173 and 184] (Table 2.2).

The pre-separation methods, precipitation of titan hydroxide, anion-exchange separating and scavenging with lead fluoride, were examined and the higher sensitivity of these methods compared with non-destructive determinations were observed[155] for the determination of Al and V. By anion-exchange chromatography in a mixed solvent system the separation of Lu, Yb and Tb for neutron-activation analysis was possible[122]. Lu and Yb were eluted in the same fraction, while Tb was obtained in a separate fraction.

In the presence of Tiron, bismuth passed through a cation-exchange column, whereas by addition of ethylenediamine Cu, Cd and Pb were sorbed on their complexes for pre-separation and determination by stripping voltammetry[169].

Table 2.2. Matrix separations

Elements	Separation method	Material	Detection	Detection limit	Ref.
Alkali-, earthmetals	Sephadex-gel	Molybdenum	Emission spectrometry, AAS	ppm-Range	[138]
Rare-earth elements	Cellulose	Uranium	Emission spectrometry	ng-Range	[137]
Metals	Reversed-phase with tri-butylphosphate	Uranium solutions	AAS	–	[159]
Na, K, Mg, Ca	DEAE-Sephadex: ion-exchange gel	Tungsten	AAS	1–20 ppm	[145]
Zn, Cd, Hg	Modified silicagel with Pb- or Ag-dithizonate	Cuprum matrix	Photometry	ng-Levels	[147]

2.2 Matrix Separations (see Table 2.2)

Molybdenum(VI) sorption was achieved on Sephadex G-10 at pH 2.5 in ammonium chloride solutions by the formation of a Mo-NH_3-glucose complex[138]. Alkali- and alkaline earth metal traces remained quantitatively in the effluent and were determinable with a coefficient of variation of 10% up to 60 ppm, and 5% at above 60 ppm. Uranium was separated from rare-earth elements using a cellulose column and an ether-nitric acid separation procedure[137].

Reversed-phase chromatography on a tri-butylphosphate-impregnated Kel-F column was used to remove uranium and other actinides from Al,Ca,Cd,Cu,Cr,Fe, Li,Mg,Na and Ni[159]. The effluent was 5 M HNO_3, saturated with tri-butylphosphate, which was evaporated to dryness for atomic absorption spectrometry.

Tungsten was separated on DEAE-Sephadex for determination of trace amounts of alkali and alkaline-earth metals by AAS without matrix interference[145]. Dithizone in o-dichlorobenzene was used as the stationary phase on a Chromosorb support for the enrichment and separation of Zn,Cd,Hg and Cu[147]. By adding selective masking agents to the sample solution the adsorption of the traces was prevented; the matrix was adsorbed. The separation of mercury from large amounts of copper on columns coated with silver dithizonate for example was reported.

2.3 Separations by Liquid Column Chromatography with Continous-Flow Detection (see Table 2.3)

For ultramicro determination of alkali and alkaline-earth metals ion-exchange chromatography was combined with a high-sensitivity hydrogen flame ionization-detector[120, 121].

Table 2.3. Separation by liquid column-chromatography with continous-flow detection

Element	Chromatography method/material	Applications: material	Kind of detection	Detection limit (detect. amount	Ref.
Alkali, alkaline-earth metals	Ion-exchange/ Zr phosphate	Synthetic Mixtures	Hydrogen flame ionization detector	Ca. 10^{-8} g (Cs)	120)
Alkali, alkaline-earth metals	Ion-exchange/ Zr phosphate	Synthetic Mixtures	Hydrogen flame ionization detector	$10^{-10}-10^{-12}$g	121)
Metal ions	Ion-exchange	–	Luminescence: addition of luminol + H_2O_2	Ca. 10^{-9} g	132)
Cations	Cation-exchange	–	Photometry: reaction with PAR	$10^{-5}-10^{-6}$ g	140)
Cations	Cation-exchange	–	Photometry: reaction with PAR or PAN	$10^{-5}-10^{-6}$ g	141)
Anions, cations	Ion-exchange	Waste streams, Waters, blood serum, urine, fruit juices	Conductivity	ppm-Range	170)
Cu, Co	Ion-exchange	–	Luminescence: reaction with luminol	Ca. 10^{-5} g	154)
Sn	Ion-exchange	NBS materials	UV-225 nm	100 ppm	130)
Zn	Ion-exchange	Natural water	Photometry: reaction with Zincon	ppb-Range	150)
Cl^-, Br^-, I^-	Ion-exchange	Sea water	Potentiometry	ppm-Levels	129)
Re	Reversed-phase: PTFA + tri-butyl-phosphate	Molybdenite concentrates, flue dust	UV-227 nm	20 μg	157)
U	Silica gel (U complex)	Rocks, underground waters	Photometry: reaction with Arsenazo III	μ-Amounts	174)

forced-flow

Metal ions	Ion-exchange, stepwise elution-technique	–	UV-225 nm	μg- To ng-Range	164) See also 140, 141)
Bi	Ion-exchange	NBS-standards	UV-254 nm	Ca. 10 ppb	187)
Fe	Ion-exchange	NBS-standards	Photom. 355 nm	10^{-2}%	166)
Pb	Ion-exchange	Standard samples	UV-270 nm μg-	To ng-Range	165)
Sb	Ion-exchange	Standard alloys, human hair	Coulometry: reaction + I_2	Ca. 1 ppm	179)
NO_3^-, NO_2^-	Ion-exchange	Cornstalks	Coulometry, see also Ref.[128)	0.1 ppm N	126)

Table 2.3. (continued)

Element	Chromatography method/material	Applications: material	Kind of detection	Detection limit (detect. amount)	Ref.
High-pressure liquid-chromatography: HPLC					
Cu, Ni, Co	Silica gel: sep. of diethyldithiocarbamates	–	UV-254 nm, argon-plasma d.	5–10 ng	183)
Cu, Ni	Silica gel, sep. as complexes with N,N′-ethylene-bis(acetyl-acetonimine	–	UV-254 nm	ng-Levels	181)
Cu, Ni, Pb, Hg, Co, Se, Cr	Reversed-phase: sep. of diethyldithiocarbamates	–	UV/VIS-spectrometry	ng-Levels	160)
Cu, Ni, Pd	Reversed-phase: sep. as complexes with β-ketoamines	–	UV-254 nm	ng-Levels	182)
Ni, Cu	Reversed-phase: sep. of tetradentate β-ketoamine chelates	–	UV-254 nm	0.2 ng Ni 0.5 ng Cu	131)
Se, NO_2^-	Reversed-phase: Se as piazselenole, NO_2^- as 2,3 naphthotriazole	Water, biolog. samples	UV, fluorimetry	0.5 ppm (Se) 0.1 ppm (NO_2^-)	186)
Se	Reversed-phase: as piazselenol or diethyl-dithiocarbamate	–	UV, fluorimetry	ng- To pg-Levels	161)
Se	Reversed-phase: chlor-piazselenol	Water samples	UV-254 nm	0.32 ppb	162)
Hg	Reversed-phase of dibenzo-18-crown-6 complexes	–	UV-254 nm	0.1 µg	148)
S	Reversed-phase with gel (styrene-divinyl-benzene)	Water, oil	UV-254 nm	Ca. 10^{-9}	123)
Cu	Ion-exchange of NTA-, EDTA-complexes	Water	UV-235 nm	ng-Range	136)
Thiosulfate, polythionates	Ion-exchange	Environmental samples, waste water	Fluorimetry: reaction with Ce(IV)	0.3 ppm	188)

The chemiluminescence reaction of metal ions with luminol and H_2O_2 was used to detect traces of twenty metal ions in the effluents after ion-exchange chromatography[132, 154].

For separations on macro- and microreticular cationexchange resins in column chromatography aqueous acetone-hydrochloric acid solutions were used and the ef-

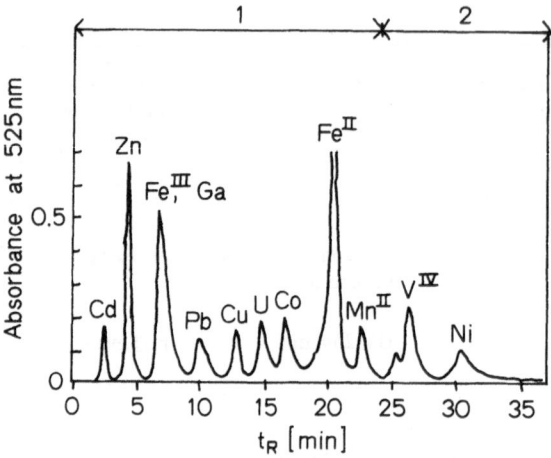

Fig. 2.1. Liquid column chromatographic separation of metal ions on Amberlite 200 (25–30 μm), stepwise elution-chromatography with hydrochloric acid-acetone (1) and hydrochloric acid-acetone-dimethylformamide mixtures (2), detection system: effluent reaction with PAR, absorbance measurement at 525 nm, amounts: 10^{-8} mole range (after Kawazu 1977, Ref.[140])

ficiencies were compared[140]. By stepwise elution chromatography trace amounts of Cd,Zn,Fe,Cu,Pb,Co,Mn,V and Ni were separated and automatically detected by flow spectrometry within 35 minutes (Fig. 2.1). The metal reagents PAR and PAN were used for reaction after separation[140, 141].

For using a conductivity cell as a universal and very sensitive monitor of ionic species, a novel combination was used to neutralize or suppress the background from eluents in the ion-exchange chromatography without significantly affecting the species being analyzed (Fig. 2.2)[170]. Automated analytical schemes are given for separations of several cations and anions. By a polyacrylate resin, XAD-11, only Sn, Hg and Au are removed from 1 M HCl solution. Tin analysis is described with contiuous-flow detection at 225 nm[130]. The selective analysis of zinc in water was possible by separation at a cation-exchange resin and measurement of absorbance in the eluent after reaction with the reagent Zincon[150].

The potentiometrical detection by a silver-silver chloride microelectrode was used for the determination of trace amounts of halides in the presence of other ions

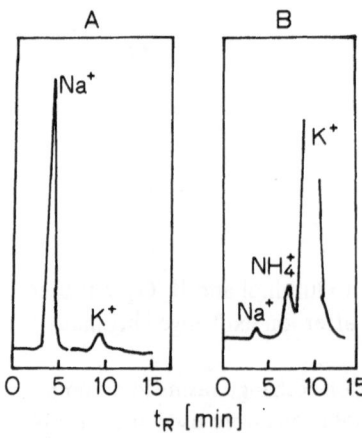

Fig. 2.2. Ion-exchange-chromatography of K^+, Na^+ and NH_4^+ from dog's blood serum (A) and grape juice (8) by detection with conductivity measurements (after Small et al. 1975, Ref.[170])

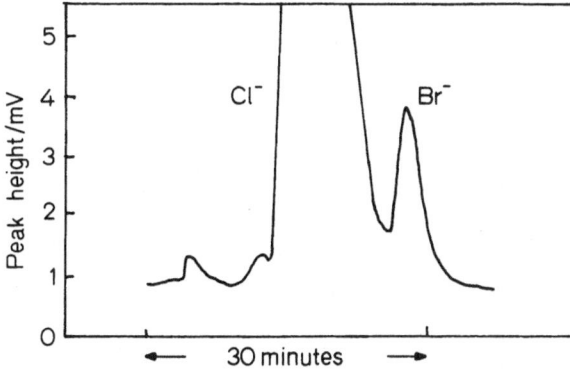

Fig. 2.3. Direct determination of bromide in sea water by liquid column-chromatography (ion-exchange) and potentiometric detection, bromide content: 60 ppm, chloride content: 20,000 ppm (after Franks and Pullen 1974, Ref.[129])

(Fig. 2.3)[129]. Ion-exchange with Zeo-Karb materials allowed the separation of chloride, bromide and iodide and their determination in sea water. Theoretical background and experimental details of the detection system are given.

For continuous-flow detection in the UV region ReO_4^- was separated from MoO_4^{2-} and other absorbing ions by reversed-phase chromatography with tri-butylphosphate on PTFA[157]. Previous equilibration with M H_2SO_4 was necessary. Interfering species were removed by washing the column with M H_2SO_4, rhenium was eluted with water. The spectrophotometrical determination of uranium traces after reaction of effluents with Arsenazo III is described[174]. From the solution containing EDTA and tartaric acid at pH 5, uranyl ions were selectively adsorbed on a silica gel column.

2.3.1 Forced-Flow Chromatography

Anion-exchange adsorption behavior of ten metal ions in mixtures of HCl and $HClO_4$ was studied[164]. By forced-flow chromatography and automatic UV detection of the eluent stream trace amounts could be determined[140, 141].

By elution with 0.5 M HBr bismuth was selectively separated from other metal ions on a cation-exchange column, the elution was spectrophotometrically and automatically recorded (Fig. 2.4)[187]. A rapid and selctive separation method for iron was described, which used a macroreticular Amberlyt A-26 resin (strong-base anion-exchanger)[166]. The iron was concentrated on a small volume and then stripped with dilute HCl. Continuous determination in the effluent followed by spectrophotometry.

Similar forced-flow chromatography separation using anion-exchange chromatography and automatic detection is described for lead[165]. With a platinum coulometric detector the highly irreversible electrochemical oxidation of Sb(III) in dilute HCl was electrocatalyzed by I^- or I_2 and specifically adsorbed at the electrode surface[179]. This detection system was used coupled with the ion-exchange separation of antimony.

After separation with a strongly basic anion-exchange resin, dissolved NO_3^- and NO_2^- were separated within nine minutes and were detected by a Cd coulometric detector[126]. A rotating Cd disk electrode was also used.

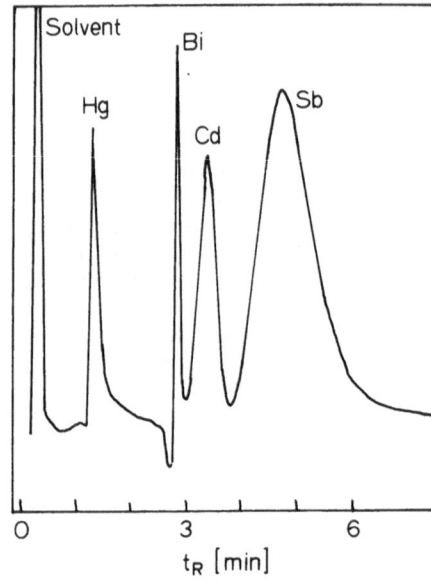

Fig. 2.4. Determination of bismuth by forced-flow liquid-chromatography, resin: Amberlyst 15, 250—325 mesh, 0.17 μg Bi(III), detection: UV-254 nm (after Willis and Fritz 1974, Ref.[187])

2.3.2 High-Pressure Liquid Chromatography (HPLC)

The metal diethyldithiocarbamates, used in gas-chromatography (Chap. 1), were also separated by adsorption high-pressure liquid chromatography on 8 μm diameter spherical silica (elements: Cu, Ni, Co)[183]. In this instance no problems such as those in gas-chromatography arose, because there was thermal instability. To confirm the metal content of eluted peaks and permit specific detection, an argon-plasma emission spectroscopic detector was used as well as a UV detector. Complexes of Cu and Ni with N-N'-ethylenebis(acetylacetoneimine) or N-N'-ethylenebis(salicylaldimine) were also separated by adsorption HPLC with UV detection[181]. The reversed-phase chromatography with chemically stable bonded stationary phases (alkyl-groups on silica gel over silicium bridges) was used for separation of some metal complexes (Table 2.3)[160, 182, 131, 148]. Applications are published for the determination of selenium in water samples (Fig. 2.5[186, 162]).

The reaction which was used was previously described for the gas chromatographic determination of compounds such as piazselenol (Chap. 1); comparable sensitivities were obtained by UV and fluorimetric detection. No difficulties in elution were observed for liquid column chromatography in comparison with gas chromatography[64].

Further applications are described for sulfur (Fig. 2.6) and copper in water by reversed-phase or ion-exchange high-performance liquid chromatography[123, 136]. Also anions such as thiosulphate and polythionates were separated from environmental samples by HPLC (anion-exchange)[188]. The sensitive detection was made possible by reaction with Ce(IV) after separation and fluorimetry of Ce(III) ions. The developments and applications of HPLC for inorganic trace analysis are still in their beginning stages.

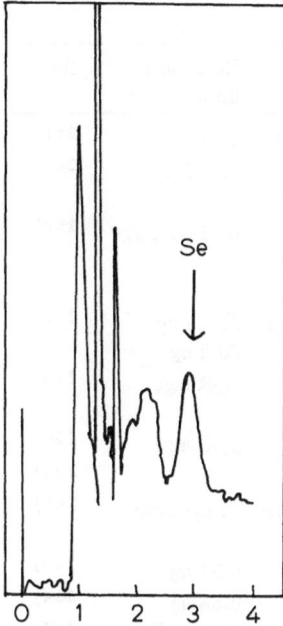

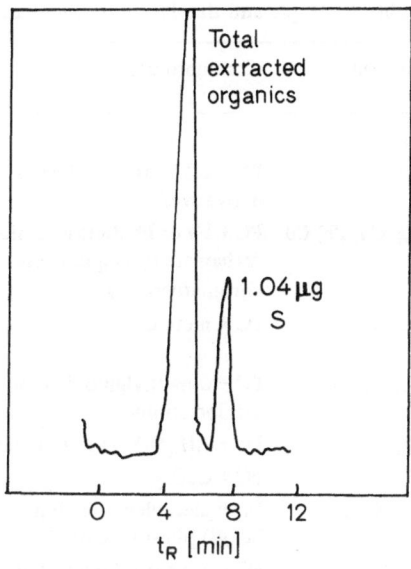

Fig. 2.5. Determination of selenium in water by reversed-phase HPLC of 5-chloropiazselenol (content: 0.8 ppb), detection: UV-320 nm (nach Schwedt and Schwarz 1978 , Ref.[162])

Fig. 2.6. Analysis of elemental sulfur by HPLC, material: process water from heavy-water plant (chloroform extract), detection: UV-254 nm (after Cassidy 1976, Ref.[123])

3 Paper and Thin-Layer Chromatography

Numerous papers deal with the separation of ions, inorganic substances and metal chelates an layers using paper and thin-layer chromatography[189–193]. However, only a few publications provide exact and complete quantitative data. These methods and techniques are important for inorganic trace analysis because they offer simplicity of separation, the use of selective spray reagents for detection and the possibility of automatization of quantitative evaluation by scanning methods.

Classic paper chromatography like distribution chromatography is also carried out today in most cases in layers on a support sheet. To obtain low standard deviations extensive automatization from sample application to evaluation is necessary. The essential advantage in comparison to liquid column chromatography is the possibility of separating several samples side by side, because therefor usually no apparatus is required.

3.1 Separations of Trace Amounts

In Table 3.1 a number of recent studies which concern themselves with the more important aspects of separation by paper and thin-layer chromatography for inorganic

Table 3.1. Paper and thin-layer chromatography of trace amounts

Element	Chromatography	Detection	Detection limit	Ref.
Cs, Rb	PC: as picrates	Elution: photometry	5 μg	251)
Cu, Co	PC: as 2-hydrazinothiazole derivatives	Planimetry	0.1–6 μg	267)
Ag, Cu, Pb, Cd	PC + Fe or Pb diethyldithio-carbamate (precipitation-chromatography)	Visual	0.05–0.2 μg	196)
Pt, Pd	PC + metallic Ag	Visual: redox process	Pt 0.5 ng, Pd 2 ng	212)
Ni, Cu	PC + dimethylglyoxime or dithiooxamide	Visual	μg-Range	213)
Zr, Hf	PC + NH_4NO_3 (ascending PC)	+ Arsenazo III	μg-Range	226)
Fe	PC + $CaCl_2$	Visual	7 μg	237)
Zr, Nb	PC + methylene-bis(di-n-hexyl) phosphine oxide	+ 8-Hydroxychinoline	Zr μg-range	260)
Co	PC + 1-nitroso-2-naphthol	Planimetry	0.31 μg	263)
Mg	PC + chinalizarine	Visual	0.68 μg	264)
Ag	PC + azorhodanine groups	Photometry	$5 \times 10^{-7}\%$ on 5 mg cellulose	284)
Te	PC + Bismuthiol II	Visual	0.15 μg	285)
Ni	PC + zinc diethylxanthogenate	Visual	2 μg	311)
S	PC + Ba rhodizonate	Planimetry	0.22 μg	204)
U (acetato complex)	DEAE(cellulose)anion-exchanger	+ $K_3[Fe(CN)_6]$ (ring-oven)	ng-Range	194)
Zn	Cellulose + liquid ion-exchanger	+ PAN (densitometry)	1 μg	241)
Zn, Cd, Co, Fe	Cellulose + liquid ion-exchanger	+ PAN (densitometry)	1 μg	242)
Au, Se, Te	Al_2O_3, Al_2O_3-glass powder, Al_2O_3-silica gel: TLC	+ $SnCl_2$	0.5 μ g	236)
Fe, Co, Mn, Ni, Zn, Co, Rh, Pb, Cd, Hg	TLC: silica gel, as tetra-phenylporphyrin chelates	Visual	10^{-10} moles	253)
P as PO_4^{3-}	TLC: silica gel	Elution: molybdato reaction	0.2 μg P	261)
Ni, Cu, Co, Mn	TLC: silica gel, as PAN-chelates	Elution: photometry	0.4 ppm	227)
Co, Cu, Ni	TLC: Al_2O_3 + dithizone	Reflectance measurement	Cu 1,2 μg, Ni 6.2 μg, Co 0.8 μg	266)
F, Cl, Br, J	TLC: silica gel + starch binder fluorescent reagent	Visual: + bromcresol-purple	1–2 μg	310)
Cu, Cu, Ni, Fe	TLC: Al_2O_3, combination of ring-oven and circular chromat.	Visual: + rubeanic acid, + KSCN	5–10 ng	233)
S-anions	Cellulose: ring-colorimetry	Different spray reagents	0.2–1 μg	245)

trace analysis are compared. In paper chromatography, for example, metals present as picrates[251] or 2-hydrazinothiazoles[267] as well as in the form of ions (e. g.[274]) are separable by distribution chromatography and they can be determined in the μg-range after elution or directly on the layer.

Metal ions were, moreover, successfully separated on impregnated paper and quantitatively registered. Organic chelating reagents (Table 3.1) or inorganic substances[212, 226, 237] are useful for impregnation. The organic group can also be bonded to the cellulose (e. g. as an azorhodanine group)[284].

Cellulose ion-exchange materials[194, 247, 248] and layers of cellulose which have been impregnated with liquid ion-exchangers[241, 242] can be used to separate ions.

In thin-layer chromatography cellulose can also be put in as a stationary phase[230, 231, 234, 235, 245, 281]. Alumina[236] and silica gel[227, 253, 261] and mixtures of them[236] can also be used. It is possible to separate ions by distribution chromatography[236, 261] and chelates by adsorption chromatography[227, 244, 253]. Impregnated layers are also used in thin-layer chromatography as well as in paper chromatography[266].

Higher detection sensitivities can be regularly obtained by the application of the ring-oven technique in connection with thin-layer chromatography[233, 245, 256, 257, 258].

3.2 Detection and Determination Methods

For quantitative analyses in paper and thin-layer chromatography the following evaluation and detection methods are in use (Table 3.2.1)[298]:
the visual method[199, 292];
planimetry — colored substances[207]
              — after spraying[290];

Table 3.2.1. Determination methods in paper and thin layer chromatography

Element	Chromatography	Detection	Detection limit	Ref.
Ru, Os	PC	Visual: + thiourea	ng-Range	292)
Sn	TLC: silica gel	Visual: + 2,2'-diquinoxalyl	20 μg	199)
Mn, Co, Ni	PC as 2-hydrazinothia-zonates	Planimetry	Mn 8 ng, Co 20 ng Ni 60 ng	207)
Cd, Bi, Pb	Ascending PC	Planimetry: + 2-mercapto-5-anilino-1,3,4-thiodiazole	Cd 30.7 μg, Bi 0.1 μg, Pb 0.4 μg	290)
Cd, Co, Ni, Zn, Hg	TLC: silica gel + diato-meous earth as dithi-zonates	Spectrophotometry	1 μg	224)
Ni, Co, Cu	PC	Photometry *in situ:* + rubeanic acid	2 μg	250)
In	PC	Photometry *in situ:* + alizarin	0.25 μg	262)

187

Table 3.2.1. (continued)

Element	Chromatography	Detection	Detection limit	Ref.
Co, Ni, Cu, Bi, PB, Mn, Cd, V, U	TLC: cellulose, silica gel	Photometry *in situ:* PAR, PAN	0.01 μg	281)
U	PC	Photodensitometry: + $K_3[Fe(CN)_6]$	10 μg	205)
Se	PC	Photodensitometry: + thiourea	3 μg	283)
Ni, Co, Cu, Pb, Mn, Cr, Hg, Zn, Bi, Fe, Ba, Sr	TLC: silica gel	Photodensitometry: colour reagents	0.1–3.2 μg	306)
Ni, Co, Cu	TLC: alumina, silica gel, cellulose	Diffuse reflectance spectrometry: + rubeanic acid	50 ng	231)
Ni, Co, Cu	TLC: alumina, silica gel, cellulose	Diffuse reflectance: + pyridine-2-aldehyde-2-quinoyl-hydrazone	10 ng	234)
Cu, Ni, Co	TLC: cellulose, silica gel as chelates	Reflectance spectrometry	10–20 ng	230)
Co, Ni, Zn	Cellulose	Reflectance spectrometry: + neocuproin or dimethylglyoxime	μg-Range	235)
Fe, V, Ti/Ge, As, Sb	Cellulose/silica gel	Fluorimetry, photometry: + tropolone/thiocarbonate resp. quercitin	0.8–4 μg	255)
Pb, Bi, U, Th, Cu, Co, Ni, Cd	TLC: fluorescent sup-	Photometry, fluorimetry: chromogenic reagents	μg-Amounts	254)
Mg, Ca, Sr	PC	Fluorimetry *in situ:* + 8-hydroxychinoline	μg-Amounts	269)
Sn	Cellulose: Chloro-complex	Fluorimetry of the chloro-complex	0.5 μg	313)
Alkaline-earth metals	PC	Electrometry	50 μg	216)
Ni	PC	Radiometry: Ni ^{35}S	10 ng	217)
SO_4^{2-}, PO_4^{3-}	Indium oxide layer (4 μm size, thin-film)	Autoradiography: ^{35}S, ^{32}P	10^{-14} g	222)
Li, K, Cs	PC	Flame photometry	μg-Range	220)
Cu, Pb, Bi	PC	+ Oxine: elution, photometry	4 ppm	200)
Ba, Ca, Mg, Sr	Cellulose	+ Oxine: elution, emission spectrometry	0.05 ppm	232)
Hg, Cu, Pd, Bi, Ni, Co, Zn, Pb	TLC: silica gel, as dithizonates	Elution: photometry	0.125–0.5 ppm	305)
Al, Be	PC	+ Salicylaldehyde, elution, spectrofluorimetry	μg-Range	203)
PO_4^{3-}	PC	Mineralization, molybdato-reaction, photometry	0.1 μmole	239)

spectrophotometry in situ, photodensitometry of colored zones[224, 307, 309, 310];
spectrophotometry in situ, photodensitometry after spraying[205, 250, 262, 273, 281, 283, 306];
diffuse reflectance spectrometry[230, 231, 234, 235, 273];
fluorimetry[254, 255, 269, 313];
electrometry[216];
radiometry (autoradiography)[217, 222];
flame photometry[220];
elution with different spectrometric methods for determination[200, 203, 232, 305];
determination after mineralizing (for paper chromatography)[239].

Quantitative details for the reproducibility of detection limits give a picture of the precision and the sensitivity of the various evaluation methods (Table 3.2.2). The data are stated either as error or as standard deviation in the literature.

Table 3.2.2. Quantitative results of paper and thin-layer chromatographic determinations (examples) (For chromatography details see Table 3.1 or 3.2)

Elements	Detection method	Detection limit	Quantitative error	Results S. D.	Ref.
Zr, Hf	+ Arsenazo III: visual	Zr: 17.7 μg – Hf: 17.1 μg	Zr 1.9% Hf 0.5%	–	226)
Te	+ Bismuthiol II: visual	0.15 μg	2%	–	285)
Pt, Pd	Planimetry: peak height	Pt 0.5 ng – Pd 2 ng	Pt: 5–9% Pd: 5–14%	– –	212)
Ni, Cu	Planimetry: peak height	Cu: 2.5 mg/ml Ni: 50 μg/ml	Cu: 10%, Ni: 50%	–	213)
Ni, Co, Cu	+ Rubenanic acid: diffuse reflectance spectr.	0.05 μg	2–5%	–	231)
Co, Cu, Ni	+ TAR: reflectance spectr.	0.01–0.02 μg	–	1.5%	230)
Several metals	Colour reagents: densitometry	0.1–3.2 μg	–	0.2–2.1%	306)
Zn	+ PAN: densitometry	1 μg	–	5%	241)
Zn, Cd, Co, Fe	+ PAN: densitometry	1 μg	–	4%	242)
Ni, Co, Cu	+ Rubeanic acid: photometry *in situ*	2 μg	3%	–	250)
Cd, Co, Ni, Zn, Hg (as dithizonates)	Elution: photometry	1 μg	Hg 10%, Cd 7.8%, Ni 12%, Zn 11%	Co 8.5% – –	224)
Heavy metal	Elution: photometry				
Heavy metal dithizonates	Elution: photometry	0.125–0.5 ppm	–	4%	305)
Ba, Ca, Mg, Sr	+ Oxine: elution, emission spectrometry	0.05 ppm	10%	–	232)
S	PC + Ba rhodizonate, planimetry	0.22 μg (–1.13 μg)	–	0.6–1.5%	204)

In the μg-range visual methods also give satisfactory results[226, 285]. Evaluations in the nanogram-range are[212, 213] possible using planimetry. Diffuse reflectance spectrometry or densitometry yield results with relative standard deviations of 0.2–5% in the nanogram- and microgram-range[230, 241, 242, 250, 306]. If an elution is carried out before photometric determination, larger mistakes appear[224, 232]. In favorable cases relative standard deviations of 4% can also be obtained[305].

A look at Table 3.2.2 shows that quantitative paper and thin-layer chromatography, above all in the microgram-range, are comparable with other chromatographic methods or techniques used in inorganic trace analysis (Table 1.3).

3.3 Applications

The significance of paper and thin-layer chromatography lies in the possibility of pre-separating single elements by simple means and also in the separation of element mixtures from different materials in connection with detection and determination methods. Disadvantages for quantitative trace analysis are reproducibility problems

Table 3.3.1. Application of paper-chromatography

Element	Chromato-graphy	Material		Ref.
Ni, Co, Cu, Zn, Nb, U	PC	Rocks	Comparisons with various other analyt. methods	195)
Cu, Co, Ni	Strip-PC	Minerals	Elution, various reagents	218)
Ni, Co, Cu, Mn	PC	Silicates	Comparison with spectrography, photometry, AAS	223)
Fe, Cu, Co, Zn, Mn	PC	Foods	Review	282)
trace elements	8-Hydroxy-chinoline-extraction, PC	Biological materials	Various methods	228)
Cations	Circular-PC	Dyes	Several spray reagents	304)
Alkali metals	Strip-PC	Minerals	+ Violuric acid, elution	202)
Alkali metals	PC	Water	Elution, flame photometry	
Cr, Ni	PC	Air	+ o-Hydroxychinolin	268)
Co, Ni, Cu	PC	Air	+ Rubeanic acid	272)
Si	PC	Biological material	Planimetry	271)
Toxicology metals	TLC: as dithizonates	Autopsy tissues	Photometry	308)
Heavy metals	TLC: as dithizonates	Water, sludge	Photometry	294)
Hg, Zn, Bi, Ni, Co, Cd	TLC: as dithizonates	Pharmaceuticals	Photometry	287)
Hg	TLC: as dithizonates	Ointments, biol. mat.	Photometry	275)

(above all in paper chromatography) in separations and in the quantitative evaluation as well as the long separation times.

For this reason, the number of quantitative applications in comparison to the described separation possibilities and qualitative analyses is relatively small. Distribution chromatographic separations at cellulose are carried out mostly at layers on plates today. In Table 3.3.1 publications are compared, which review the applications of paper chromatography to various materials such as minerals, food, biological

Table 3.3.2. Quantitative results of paper and thin-layer chromatography: inorganic materials

Element	Chromatography	Material	Detection method	Detection limit	Ref.
Cu	Paper-chromatography	Cadmium selenide	+ Diethyldithiocarbamate	0.05%	265)
Nb	Paper-chromatography	Ores, minerals	+ Tannin	50 ng	225)
U	Paper-chromatography	Ores, water	Fluorimetry	1 ng	243)
U	Paper-chromatography	Geochemical samples	+ PAN	1 ppm	289)
Cu, Al	Paper-chromatography	Corrosion products	+ Aluminon, AAS	$0.01-0.1\ \mu g$	303)
Ru, Os	Paper-chromatography	Mineral raw materials	+ Thiourea	$1-3 \times 10^{-5}\%$	293)
Se, Te	Paper-chromatography	Minerals	+ $SnCl_2$, reflectance spectrometry	$10^{-4}\%$	238)
Se, Te	Paper-chromatography	Sulfide minerals	Elution, photometry	μg-Range	288)
Co, Cu, Ni, Zn	Paper-chromatography	Rocks, soils	+ Various reagents	5 ng Zn, 20 ng Co, Cu, Ni	270)
Ni, Co, Cu, Fe	Paper-chromatography	Sulfide ores, rocks	+ Rubeanic acid	$2 \times 10^{-4}\%$	316)
Li	TLC: cellulose	Alkali metal solutions	Fluorimetry, photometry	1 ppm	277)
Pb	TLC: cellulose	Ceramics	+ Diphenylcarbazide	$20\ \mu g$	197)
Ni, Co, Mn, Cr, Fe	TLC: silica gel	Cosmic dust	+ Rubeanic acid, densitometry	μg-Range	296)
Fe, Cu, Hg, Zn, Ni, Ca	TLC: silica gel	Rare-earth preparations	+ Rubeanic acid	$10^{-2}\%$	315)
Rare-earth	TLC: silica gel	Uranyl nitrate	+ Arsenazo III	$10^{-3}\%$	317)
Rare-earth	TLC: silica gel + bis(2-ethylhexyl)-phosphoric acid	Magnetic alloys	Various methods	μg-Range	300)
Rare-earth	TLC: silanizated silica gel	Ores, technical products, pure elements	+ Neothorine	$50-100$ ng	252, 259)

samples, air and water, as well as pharmaceuticals. Comparison of the results with those of other analytical methods were carried out. Thin-layer chromatography, above all has been applied in the separation of metal dithizonates. Examples are given in Ref.[275, 287, 294, 308]. However, these examples provide no information about the detection limit.

In Table 3.3.2 publications concerning the application of paper and thin-layer chromatography for elemental trace analysis in inorganic materials are compared insofar as they include data about the detection limit.

Not only pre-separation of some elements (Cu, Nb, U) but also separations of mixtures are described. The detection limit is given as either the content level in the material for the procedure or as the absolute amount at the layer.

In Table 3.3.3 applications for the determination of elements in organic and environmental materials are presented. Here also only those works were cited, which contain quantitative or at least semi-quantitative results.

In addition to distribution chromatographic of ions on paper by utilization of various spray reagents to help detection, thin-layer chromatography of metal chromatography of metal chelates, more than any other technique, has been widely applied with metal dithizonates in particular[297, 198, 319, 201, 321].

The kind of evaluation made determines whether a semi-quantitative or quantitative procedure will be used for the inorganic trace analysis.

Table 3.3.3. Quantitative results of paper and thin-layer chromatography: organic and environmental materials

Element	Chromatography	Material	Detection method	Detection limit	Ref.
Cu	Paper chromatography	Pharmaceuticals	+ Dithizone, elution	4 µg/sample	302)
Sr	Ring-PC	Vegetable, sea-water	+ Na rhodizonate, ashing, extrative-photometry	µg-Amounts	208)
Pb	Paper chromatography	Vegetables	Ashing, + PAR photometry	µg-Amounts	209)
Tl	Paper chromatography	Organic materials	Ashing, extractive-photom.	µg-Amounts	211)
U	Paper chromatography	Soils, plants	–	ppm-Range	291)
V	Paper chromatography	Biological materials	Decomposition, photom.	µg-Amounts	210)
Zn	Paper chromatography	Blood serum	+ 8 Hydroxyquinoline	µg-Amounts/ml	215)
Zn	Paper chromatography	Blood	Ashing, spectrometry	2.5 µg	317)

Table 3.3.3. (continued)

Element	Chromatography	Material	Detection method	Detection limit	Ref.
Zn	Paper chromatography	Urine	Elution, complexometry	3 µg	295)
Cu, Zn, Fe, Mn	Paper chromatography	Plant-leaf material	Various reagents, reflectance spectrometry	1 µg (ppm-Range)	318)
Cu, Co, Ni, Mo, Mn	Paper chromatography	Foodstuffs	Reflectance spectrometry	0.05 µg (ppb To ppm-Range)	219)
Ni, Mn, Co, Cu, Fe, Zn, Mo, V	Paper chromatography	Vegetable oils	Various reagents, photometry	1 µg	240)
Ag	TLC: silica gel	Drugs	Photodensitometry	50 ppm	276)
B	TLC: of B-curcumin complex	Caviar	Photometry	1 ng	214)
Cd	TLC as dithizonate	Pharmaceuticals	Reflectance photom.	µg-Range	279)
Mg	TLC: Al_2O_3	Cotton materials	+ Benzidine	0.1–0.3 ppm	299)
Hg	TLC as dithizonate	Foods	Elution, photometry	µg-Range	198)
Mn	TLC as PAN-chelate	Pharmaceuticals	Photometry	0.5 ppm	278)
Pb	TLC: cellulose	Mineral oil	Fluorimetry as $PbCl_4^{2-}$	2 ppb	314)
Pb	TLC: cellulose	Oils	Fluorimetry as $PbCl_4^{2-}$	0.3 µg	312)
Pb	TLC	Air	+ Dinatrium-tetra-Hydroxyquinoline	µg-Amounts	246)
Pb	TLC	Blood	Elution, photom.	20 µg/100 ml	206)
S	TLC	Shampoo	Spectrodensitometry	0.3 µg	286)
Al, Be, Cr	TLC: silica gel	Water	Fluorimetry, photom.	Al 0.1 ppm Be, Cr 1 ppm	320)
Co, Cu, Ni	TLC: silica gel	Ceveals	+ Rubeanic acid, photometry in situ	ppb-To ppm-levels	229)
Hg, Bi, Cu, Zn, Sn	TLC as dithizonates	Water	Photometry	0.5 ppm	319)
Ag, Cd, Co, Cu, Hg, Ni, Pb, Zn	TLC as dithizonates	Blood, urine, excrement	Elution, photom.	100 ng/l	201)
Zn, Cu, Ni, Pb, Hg, Cd	TLC as diethyl-dithiocarbamates	Water	+ Dithizone	ng-Range	297)
14 Metals	TLC as dithizonates, oxinates	Water	Photometry	0.1 ppm: Fe. Cu, Hg, Cd 0.5 ppm: Cu, Ni, Mn, Pb, Zn, Bi, Sn 1 ppm: Al, Be 8 ppm: Cr	321)
Alkaline e	TLC: cellulose	Vegetable ash	+ 8-Hydroxychinoline, fluorimetry	Ba 50 ng, Sr 20 ng, Ca 10 ng, Mg 5 ng	280)

4 Chromatographic Methods for Enrichment
   (see Table 4)

Numerous of the column chromatographic methods described in this paper make possible or simultaneously included the enrichment of the separated elements (Chaps. 2.1 and 2.2). In this section only those publications are mentioned which concentrate on enrichment applications.

Ion-exchange procedures in particular are suited for the enrichment of cations and anions. Examples for anion-exchange concentration of impurities in rare-earth materials are given[337]. Anion-exchange is also used for the enrichment of the uranyl thiocyanide complex[328]. Uranium is directly determined on the resin by delayed-neutron counting. Complexes of cobalt, zinc and cadmium were enriched on an anion-exchanger. A ligand-loaded resin column can also be used to remove them (Table 4.1)[331]. In addition to the conventionally used resins like Dowex polyurethane resins[341] and carboxycellulose exchangers are also described[342, 343]. Enrichment by resins is comparable with co-precipitation methods (with PbS, $Fe(OH)_3$ $Al(OH)_3$) in the sorption efficiency[341]. Frequently chelating resins are described for the enrichment of several heavy metals[329, 333, 334, 340]. In the simplest form the chelating agent is immobilized on a glass support[332]. A comparison was made between anion-exchange and co-precipitetion, which showed the same performance in the recovery of inorganic traces but more efficient[332]. Retention of cadmium, cobalt, copper, nickel and zinc by a column of Ca-Chelex is achieved at pH 6.5[329]. At lower pH's slow resin kinetics lead to trace metal losses. The fundamental effects of this have been studied. A mixture of a Chelex 100 resin and Pyrex glass powder of the same mesh-size has been successfully used for enrichment of several elements[334]. Furthermore chelating resins were produced by co-polymerisation of styrene and 3(5)-methylpyrazole[324] or by diazo-coupling of dithizone to form a modified carboxymethylcellulose[325]. The first chelate sorbent was boiled with a solution of noble metals. Complex formation was indicated by IR-spectrometry. After filtration the resin was then calcinated with a mixture of perchloric, nitric and sulfuric acid[324]. For the pre-concentration, partial separation and quantitative recovery of trace elements from sea water some ligands such as dithizone or oxine, diazo-coupled ad the second ion-exchange material, were used in a column- or also sheet-form-technique[325].

The simultaneous pre-concentration of manganese, chromium and vanadium is possible by combination of ion-exchange with redox reactions on a Fe(II)-treated resin[335]. Beside ion-exchange chromatography reversed-phase methods are very important for trace enrichments[322, 323, 338]. On inert materials (e. g. PTFA, Teflon) tri-butylphosphate (TBP) is used as stationary phase. This method is also practiced by TBP-loaded polyurethane foam[327]. After extraction of aurum(III) from acidic solutions of thiourea into TBP the complex is separated and enriched in such foam columns. Plasticized foam with dithizone allows the enrichment of silver[326]. The effects of different plasticizers, pH of the aqueous phase and dithizone concentration in the foam on the collection rate of silver were investigated. Furthermore activated carbon was used for pre-concentration of several cations[344] with a high

Table 4. Trace enrichment by chromatographic methods

Elements	Chromatographic system	Material	Concentration factor or (detection limit)	Ref.
U	Anion-exchange: uranyl thiocyanide complex	Water	(0.04 μg/l)	328)
Co, Zn, Cd	Anion-exchange: complexes with 2-(3'-sulfobenzoyl)-pyridine-2-pyridylhydrazone	–	300	331)
Au	Ion-exchange: Dowex 1-X 8, polyurethane resin	–	–	341)
24 metals	Carboxycellulose cation-exchanger	–	(1–100 μg/l)	342, 343)
Cu	Anion-exchanger or 8-hydroxychinoline immobilized on a glass support	Natural waters	(μg-Range/l)	332)
Cd, Co, Cu, Ni, Zn	Chelex 100 (Ca^{2+}-form)	Natural waters	(μg-Range/l)	329)
Heavy metals	Chelex 100 (Ca^{2+}-form)	Natural water, snow, limestone, biol. materials	–	333)
Ba, Ca, Cd, Ce, Co, Cr, Cu, Fe, La, Mg, Mn, Sc, U, V, Zn	Chelex 100	Water samples	–	334)
Noble metals	Copolymer of styrene and 3(5)-methylpyrazole: TLC	–	–	324)
Pb, Zn, Cd, Mn(II)	Dithiozone (or oxine) diazo-coupled to a modified carboxymethylcellulose	Sea water	10^5 To 10^8	325)
MnO_4^-, CrO_4^-, VO_3^-	Reactive adsorption on Fe(II)-treated resin	–	40	335)
Ga	Reversed-phase: PTFA + TBP	–	–	323)
Au	Reversed-phase: fluoroplast + TBP	Au-alloys	–	338)
Au	Reversed-phase: TBP-loaded polyurethane foam	–	–	327)
Ag	Plasticized foam containing dithizone	–	(0.01 μg/l)	326)
Several cations	Activated carbon	Water	10^4	344)
Several cations	Chelex 100 ion-exchange membrane	Water	(μ-Range/l) 1250	345)
Pt-group, Au	Ion-exchange + paper chromatography	–		339)
Active and inactive traces elements	Modified paper chromatographic round filter procedure	–	10^4	336)
Co, permanent gases	Molecular sieve	Air	(0.01 ppm)	330)

concentration factor (10^4). This is possible by combination of multielement chelation by 8-hydroxychinoline with subsequent adsorption.

Among these column techniques ion-exchange membranes[345], the combination of ion-exchange with paper chromatography[339] and paper chromatography found application in trace enrichment. For concentrating permanent gases (e.g. carbon monoxide) a stepwise technique was described[330]. After pre-concentration ad a molecular sieve determination follows by gas-chromatography with catalytic conversion of CO into methane for detection with a FID[30].

Table 4 shows examples of the most important different chromatographic methods for trace enrichment; the list contains only a selection of the numerous publications in this field.

5 References

1. (R) Rüssel, H., Tölg, G.: Gas-chromatography of inorganic compounds. Topics Curr. Chem. *33*, 1 (1972) – 629 cit.
2. (R) Rezl, V., Janak, J.: Elemental analysis by gas-chromatography. J. Chromatogr. *81*, 233 (1973) – 175 cit.
3. (R) Rodriguez-Vazquez, J. A.: Quantitative inorganic analysis by gas-chromatography. Anal. Chim. Acta *73*, 1 (1974) – 190 cit.
4. (R) Cram, S. P., Risby, T. H.: Gas chromatography. Anal. Chem. *50*, 213R (particular 231/232R) (1978). Cram, S. P., Juvet, R. S. jr.: Gas-chromatography. Anal. Chem. *48*, 411R (particular 428/429R) (1976)
5. Addison, R. F., Ackmann, R. G.: Direct determination of elemental phosphorous by gas-liquid chromatography. J. Chromatogr. *47*, 421 (1970)
6. Aglinlov, N. K., Zueva, M. V., Feshchenko, I. A., Faerman, V. I.: Gas-chromatographic and mass-spectrometric analysis of silicon tetrabromide. Zh. analit. Khim. *30*, 1733 (1975), Anal. Abstr. *31*, 2 B 114 (1976)
7. Akiba, M., Shimoishi, Y., Toei, K.: The gas-chromatographic determination of selenium in steel with 4-chloro-1,2-diamino-benzene. Analyst *100*, 648 (1975)
8. Akiba, M., Shimoishi, Y., Toei, K.: Gas-chromatographic determination of selenium in pure elemental arsenic and arsenic(III)oxide with 4-nitro-1,2-diamino-benzene. Analyst *101*, 644 (1976)
9. Albert, D. K.: Comparison of electron capture and hydrogen flame detectors for gas-chromatographic determination of trace amounts of metal chelates. Anal. Chem. *36*, 2034 (1964)
10. Archer, A. W.: A gas-chromatographic method for the determination of increased bromide concentration in blood. Analyst *97*, 428 (1972)
11. Bächmann, K., Buttner, K., Rudolph, J.: Increase of sensitivity in gas-chromatography by applying the principle of radiochemical amplification. Z. Anal. Chem. *282*, 189 (1976)
12. Ballschmiter, K.: Inorganic trace analysis combining thin-layer and gas-chromatography: Indirect determination of ruthenium as its thiosemicarbazide chelate, J. Chromatogr. *8*, 496 (1970)
13. Barrat, R. S., Belcher, R., Stephen, W. I., Uden, P. C.: The determination of traces of nickel by gas-liquid chromatography. Anal. Chim. Acta *59*, 59 (1972)
14. Becker, J. H., Chavallier, J., Spitz, J.: Determination of tin in zircaloy by gas-chromatography. Z. Anal. Chem. *247*, 301 (1969)
15. Becker, R., Buchtela, K., Grass, F., Kittl, R., Müller, G.: Separation and radiometric determination of small amounts of rare-earth elements in extraterrestical materials. Z. Anal. Chem. *274*, 1 (1975)

16. Belcher, R., Henderson, D. E., Kamalizad, A., Martin, R. J.: Gas-chromatography of divalent transition metal chelates. Anal. Chem. *45*, 1197 (1973)

17. Belcher, R., Majer, J. R., Stephen, W. I., Thomson, I. J., Uden, P. C.: The gas chromatography, termal analysis and mass spectrometry of fluorinated lead β-diketonates. Determination of traces of lead by the integrated ion current technique. Anal. Chim. Acta *50*, 423 (1970)

18. Belcher, R., Majer, J. R., Stephen, W. I., Thomson, I. J., Uden, P. C.: A gas-chromatographic method for the determination of low concentration of chloride ion. Anal. Chim. Acta *57*, 73 (1971)

19. Black, M. S., Sievers, R. E.: Determination of chromium in human blood serum by gas-chromatography with a microwave-excited emission detector. Anal. Chem. *48*, 1872 (1976)

20. Bock, R., Monerjan, A.: Separation and gas-chromatographic determination of monovalent thallium as the cyclopentadienyl compound. Z. Anal. Chem. *235*, 317 (1968)

21. Bock, R., Semmler, H.-J.: Separation and determination of fluoride ions with the aid of silicium organic compounds. Z. Anal. Chem. *230*, 161 (1967)

22. Bock, R., Strecker, S.: Separation and gas-chromatographic determination of traces of fluoride. Z. Anal. Chem. *266*, 110 (1973)

23. Booth, G. H., Darby, W. J.: Determination by gas-liquid chromatography of physiological levels of chromium in biological tissues. Anal. Chem. *43*, 831 (1971)

24. Bros, E., Lasa, J., Kilarska, M.: Application of helium detector in the determination of CO_2, CO, CH_4, COS and H_2S by gas-chromatography. Chem. Anal. (Warsaw) *19*, 1003 (1974), Anal. Abstr. *28*, 5 B 21 (1975)

25. Bruner, F., Ciccioli, P., DiNardo, F.: Further developments in the determination of sulfur compounds in air by gas-chromatography. Anal. Chem. *47*, 141 (1975)

26. Buchtela, K., Grass, F., Müller, G.: Radio gas-chromatography of metal chelates. J. Chromatogr. *103*, 141 (1975)

27. Burgett, C. A., Fritz, J. S.: Separation and quantitative determination of the yttrium group lanthanides by gas-liquid chromatography. Anal. Chem. *44*, 1738 (1972)

28. Burgett, C. A., Fritz, J. S.: Separation and quantitative determination of the cerium group lanthanides by gas-liquid chromatography. Talanta *20*, 363 (1973)

29. Burgett, C.: The gas-chromatography of selenium as the trimethylsilyl derivative. Anal. Lett. *7*, 799 (1974)

30. Cappon, C. J., Smith, J. C.: Gas-chromatographic determination of inorganic mercury and organomercurials in biological materials. Anal. Chem. *49*, 365 (1977)

31. Chau, Y. K., Wong, P. T. S., Goulden, P. D.: Gas-chromatography-atomic absorption method for the determination of dimethylselenide and dimethyldiselenide. Anal. Chem. *47*, 2279 (1975)

32. Cropper, E., Puttman, N. A.: The gas-chromatographic determination of fluoride in dental creams. J. Soc. Cosmetic Chemists *21*, 533 (1970)

33. Dagnall, R. M., West, T. S., Whitehead, P.: The determination of volatile metal chelates by using a microwave-excited emission detector. Analyst *98*, 647 (1973)

34. Daughtrey, E. H., Fitchett, A. W., Mushak, P.: Quantitative measurements of inorganic and methylarsenicals by gas-liquid chromatography. Anal. Chim. Acta *79*, 199 (1975)

35. DeSouza, T. L. C., Bhatia, S. P.: Development of calibration systems for measuring total reduced sulfur and sulfur dioxide in ambient concentrations in the parts per billion range. Anal. Chem. *48*, 2234 (1976)

36. Eisentraut, K. J., Griest, J., Sievers, R. E.: Ultratrace analysis for beryllium in terrestral meteoric and Apollo 11 and 12 lunar samples using electron capture gas-chromatography. Anal. Chem. *43*, 2003 (1971)

37. Flinn, C. G., Aue, W. A.: Photometric detection of selenium compounds for gas chromatography. J. Chromatogr. *153*, 49 (1978)

38. Florence, T. M., Farrar, Y. J., Dale, L. S., Battley, G. E.: Beryllium content of NBS standard reference Orchard Leaves. Anal. Chem. *46*,, 1874–1876 (1974)

39. Fontaine, R., Santoni, B., Pommier, C., Guichon, G.: Utilisation of β-diketone chelates for determination of uranium and thorium. Anal. Chim. Acta *62*, 337 (1972)

40. Foreman, J. K., Gough, T. A., Walker, E. A.: The determination of traces of beryllium in human and rat urine samples by gas-chromatography. Analyst *95*, 797 (1970)
41. Frame, G. M., Ford, R. E., Scribner, W. G., Cturtnicek, T.: Trace determination of beryllium oxide in biological samples by electron-capture gas-chromatography. Anal. Chem. *46*, 534 (1974)
42. Fresen, J.-A., Cox, F. H., Witter, M. J.: Determination of fluoride in biological materials by means of gas-chromatography. Pharm. Weekblad Ned. *103*, 909 (1968), Anal. Abstr. *17*, 3619 (1969)
43. Genty, C., Honin, C., Malherbe, D., Schott, R.: Determination of trace quantities of aluminium and chromium in uranium by gas phase chromatography. Anal. Chem. *43*, 235 (1971)
44. Glover, D. J., Hoffsommer, J. C.: Gas-chromatographic analysis of nitrate and nitrite ions in microgram quantities by conversion to nitrobenzene. J. Chromatogr. *94*, 334 (1974)
45. Gosink, T. A.: Rapid simultaneous determination of picogram quantities of aluminium and chromium from water by gas phase chromatography. Anal. Chem. *47*, 165 (1975)
46. Grys, S.: The gas-liquid chromatographic determination of inorganic iodine, iodide and tightly bound iodine in milk. J. Chromatogr. *100*, 43 (1974)
47. Hansen, L. C., Scribner, W. G., Gilbert, T. W., Sievers, R. E.: Rapid analysis for subnanogram amounts of chromium in blood and plasma using electron capture gas-chromatography. Anal. Chem. *43*, 349 (1971)
48. Hasty, R. A.: A gas-chromatographic method for the microdetermination of iodine, Mikrochim. Acta 348 (1971)
49. Hill, R. D., Gesser, H.: Flame photometric detection of metal chelates separated by gas-chromatography. J. Gas Chromatogr. *1*, No. 12, 14 (1963)
50. Holland, R. V., Board, P. W.: Determination of ozone by gas-chromatography. Analyst *101*, 887 (1976)
51. Jatridis, B., Parssaki, G.: Chlorination and gas-chromatographic determination of arsenic in oxides, sulphides, ores and alloys. J. Chromatogr. *122*, 505 (1976)
52. Jacquelot, P., Thomas, G.: Gas-chromatography of mixed complexes of nickel(II) or cobalt(II) with trifluoroacetylacetone and dimethylformamide, Bull. Soc. Chim. 702 (1971)
53. Jacquelot, P., Thomas, G.: Gas-chromatography of vanadyl trifluoracetylacetonate. Bull. Soc. Chim. 3167 (1970)
54. Jones, P., Nickless, G.: The stimation of inorganic mercury at low concentration by GC. J. Chromatogr. *76*, 285 (1973)
55. Jones, P., Nickless, G.: Determination of inorganic mercury by gas-liquid chromatography. J. Chromatogr. *89*, 201 (1974)
56. Juvett, R. S., Durbin, R. P.: Characterization of flame photometric detector for gas-chromatography. Anal. Chem. *38*, 565 (1966)
57. Kaiser, G., Grallath, E., Tschöpel, P., Tölg, G.: Contribution to the optimization of the chelate gas-chromatographic determination of beryllium in limited amounts of organic materials. Z. Anal. Chem. *259*, 257 (1972)
58. Kawaguchi, H., Sakamoto, T., Mizuike, A.: Emission spectrometric detection of metal chelates separated by gas-chromatography. Talanta *20*, 321 (1973)
59. Kowalski, B. R., Isenhour, T. L., Sievers, R. E.: Ultra-trace mass spectrometric metal analysis using heptafluorodimethyloctanedione chelates. Anal. Chem. *41*, 998 (1969)
60. Kruse, R., Rüssel, H. A.: Gas-chromatographic determination of arsenic as trimethylarsenate in biological material. Z. Anal. Chem. *286*, 226 (1977)
61. Lee, M.-L., Burrell, D.: Soluble aluminium in marine and fresh water by gas-liquid chromatography. Anal. Chim. Acta *66*, 245 (1973)
62. Luckow, V., Rüssel, H. A.: Gas-chromatographic determination of trace amounts of inorganic mercury. J. Chromatogr. *150*, 187 (1978)
63. Matthews, D. R., Shults, W. D., Guerin, M. R.: Extraction-derivatization gas-chromatographic determination of trace phosphate in aqueous media. Anal. Chem. *43*, 1582 (1971)
64. Meyer, A., Grallath, E., Kaiser, G., Tölg, G.: An extremely sensitive method of determination of selenium by gas-chromatography after evaporation in an oxygen stream, I. determination in high-purity copper. Z. Anal. Chem. *281*, 201 (1976)

65. Miyazaki, M., Kaneko, H.: Feasibility of gas-chromatography for ultramicro determination of aluminium in biological materials. Chem. Pharm. Bull. *18*, 1933 (1970); Anal. Abstr. *21*, 1290 (1971)

66. Mushak, P., Tibebetts, F. E., Zarnegar, P., Fisher, G. B.: Perhalobenzenesulfinates as reagents in the determination of inorganic mercury in various media by gas-liquid chromatography. J. Chromatogr. *87*, 215 (1973)

67. Myreson, A. L., Chludzinski, J. J.: A gas-chromatographic determination of HCN in combustion effluents in the low ppm range. J. Chromatogr. Sci. *13*, 554 (1975)

68. Nakashima, S., Toei, K.: Determination of ultramicro amounts of selenium by gas-chromatography. Talanta *15*, 1475 (1968)

69. Nota, G., Palombari, R.: Determination of cyanides and thiocyanates in water by gas-chromatography. J. Chromatogr. *84*, 37 (1973)

70. Nota, G., Palombari, R., Improta, C.: Determination of complex cyanides in water by gas-chromatography. J. Chromatogr. *123*, 411 (1976)

71. Noweir, M. H., Cholak, J.: Gas-chromatographic determination of beryllium in biological materials and air. Environm. Sci. Technol. *3*, 927 (1969). Ref. in Z. Anal. Chem. *254*, 163 (1971)

72. O'Brien, T. P., O'Laughlin, J. W.: Gas-chromatography of ternary complexes of manganes(II); iron(II), cobalt(II) and nickel(II) with hexafluoroacetylacetone and di-n-butylsulphoxide. Talanta *23*, 805 (1976)

73. O'Laughlin, J. W., O'Brien, T. P.: Synergic extraction of zinc, cadmium and lead with hexafluoroacetylacetone and di-n-butylsulphoxide and tri-n-butylphosphate and the gas-chromatography of the zinc adduct. Talanta *22*, 587 (1975)

74. Poole, C. F., Evans, N. J., Wibberley, D. G.: Determination of selenium in biological samples by gas-liquid chromatography with electron-capture detection. J. Chromatogr. *136*, 73 (1977)

75. Radecki, A., Halkiewicz, J., Grzybowski, J., Lamparczyk, H.: Gas-liquid chromatographic determination of zinc, copper and nickel in marine bottom sediments. J. Chromatogr. *151*, 259 (1978)

76. Ross, W. D., Wheeler, G.: Quantitative determination of chromium(III) hexafluoroacetylacetonate by gas-chromatography. Anal. Chem. *36*, 266 (1964)

77. Ross, W. D., Sievers, R. E.: Ultra trace analysis of beryllium by gas-chromatography, Gas-Chromatogr. Symp. Littlewood, A. B. (ed.). Rome: 1966, Ref. Gas-Chromatogr. Abstr. *55* (1968)

78. Ross, W. D., Sievers, R. E.: Rapid ultra-trace determination of beryllium by gas-chromatography. Talanta *15*, 87 (1968)

79. Ross, W. D., Buttler, G. W., Duffy, T. G., Rehg, W. R., Wininger, M. T., Sievers, R. E.: Analysis of aqueous nitrates and nitrites and gaseous oxides of nitrogen by electron capture gas-chromatography. J. Chromatogr. *112*, 719 (1975)

80. Ross, W. R., Sievers, R. E.: Quantitative analysis for trace chromium in ferrous alloys by electron capture gas-chromatography. Anal. Chem. *41*, 1109 (1969)

81. Ross, W. D., Sievers, R. E., Wheeler, G.: Quantitative ultratrace analysis of mixtures of metal chelates by gas-chromatography. Anal. Chem. *37*, 598 (1965)

82. Rudnevskii, N. K., Vyakhirev, D. A., Demarin, V. T., Zueva, M. V., Lukyanova, A. I.: Atomic absorption spectroscopy as detector in gas-chromatography (in the determination of antimony and gallium). Dokl. Akad. Nauk SSR *223*, 887 (1975); Anal. Abstr. *30*, 3865 (1976)

83. Rüssel, H. A.: Gas-chromatographic determination of fluoride in organic materials and body fluids. Z. Anal. Chem. *252*, 143 (1970)

84. Ryan, T. R., Hastings Vogt, C. R.: Determination of physiological levels of Cr(III) in urine by gas-chromatography. J. Chromatogr. *130*, 346 (1977)

85. Sakamoto, T., Kawaguchi, H., Mizuike, A.: Determination of traces of copper and aluminium in zinc by gas-chromatography with the microwave plasma detector. J. Chromatogr. *121*, 383 (1976)

86. Sakano, T., Hori, Y., Tomari, Y.: Gas-chromatographic determination of trace amounts of water in organic solvents containing active chlorine and hydrogen chloride. J. Chromatogr. Sci. *14*, 501 (1976)

87. Savory, J., Mushak, P., Sunderman, F. W., Estes, R. H., Roszel, N. O.: Gas-chromatographic micro determination of chromium in biological materials. Anal. Chem. *42*, 294 (1970)
88. Savory, J., Mushak, P., Sunderman, F. W.: Gas-chromatographic determination of chromium in serum. J. Chromatogr. Sci. *7*, 674 (1969)
89. Sazonov, M. L., Alymova, T. E., Selenkina, M. S., Zhuhovitskii, A. A.: Gas-chromatographic determination of germanium in coal. Khim. Tverd. Topl. *3*, 64 (1968); Anal. Abstr. *881* (1969)
90. Shimoishi, Y., Toei, K.: Gas-chromatographic determination of the ultramicro amounts of selenium in pure sulphuric acid. Talanta *17*, 165 (1970)
91. Shimoishi, Y.: The gas-chromatographic determination of selenium IV and total selenium in milk, milk products and albumin with 1,2-diamino-4-nitrobenzene. Analyst *101*, 298 (1976)
92. Sie, S. T., Bleumer, J. P. A., Rijnders, G. W. A.: Gas-chromatographic separation of inorganic chlorides and its application to metal analysis, II. Determination of silicon in iron and steel. Separ. Sci. *2*, 645 (1967)
93. Sieck, R., Richard, J. J., Iversen, K., Banks, C. V.: Determination of uranyl and thorium (IV) by gas-chromatography of volatile mixed-ligand complexes. Anal. Chem. *43*, 913 (1971)
94. Sievers, R. E., Eisentraut, K. J. et al.: Microanalysis by gas-chromatography: Determination of metals in lunar dust and rock, VI. Intern. Sympos. Microchem. Graz 1970, Vol. B 247
95. Sievers, R. E., Connolly, J. W., Ross, W. D.: Metal analysis by gas chromatography of chelates of heptafluorodimethylactanedione. J. Gas-Chromatogr. *5*, 242 (1967)
96. Schuphan, I., Ballschmiter, K., Tölg, H.: Polychlorinated compounds and their application to gas-chromatographic determination of elements in the ng-range, I. Polychlorinated xanthogenates and their application for the determination of nickel. Z. Anal. Chem. *255*, 116 (1971)
97. Schwarberg, J. E., Moshier, R. W.: Feasibility of gas-liquid chromatography for quantitative determination of Al(III), In(II), Ga(III) and Be(II) trifluoroacetylacetonates. Talanta *11*, 1213 (1964)
98. Schwedt, G., Rüssel, H. A.: Gas-chromatographic determination of arsenic as triphenylarsane. Chromatographia *5*, 242 (1972)
99. Schwedt, G.: Gas-chromatographic determination of traces of arsenic, Thesis Hannover 1971
100. Sowinski, E. J., Suffet, I. U.: Gas-chromatographic detection and confirmation of volatile boron hydrides at trace levels. Anal. Chem. *46*, 1218 (1974)
101. Taitiro, F., Tooru, K., Shigeo, M.: Gas-chromatography of metal chelates with carrier gas-containing ligand vapour. Talanta *18*, 429 (1971)
102. Talmi, Y., Norvell, V. E.: Determination of arsenic and antimony in environmental samples using gas-chromatography with a microwave emission spectrometric system. Anal. Chem. *47*, 1510 (1975)
103. Talmi, Y.: The rapid sub-picogram determination of volatile organo-mercury compounds by gas-chromatography with a microwave emission spectrometric detector system. Anal. Chim. Acta *74*, 107 (1975)
104. Talmi, Y., Andren, A. W.: Determination of selenium in environmental samples using GC with a microwave emission spectrometric detector system. Anal. Chem. *46*, 2122 (1974)
105. Tavlaridis, A., Neeb, R.: Gas-chromatographic multielement analysis using simple and fluorinated diethyldithiocarbamates. Z. Anal. Chem. *282*, 17 (1976)
106. Tavlaridis, A., Neeb, R.: Gas-chromatography of some bis(diethyldithiocarbamato)chelates. Naturwiss. *63*, 146 (1976)
107. Taylor, M. L., Arnold, E. J.: Ultratrace determination of metals in biological specimen – quantitative determination of beryllium by gas-chromatography. Anal. Chem. *43*, 1328 (1971)
108. Taylor, M. L., Arnold, E. L., Sievers, R. E.: Rapid microanalysis for beryllium in biological fluids by gas-chromatography. Anal. Lett. *1*, 735 (1968)
109. Tesavrik, K., Krejci, M.: Chromatographic determination of carbon monoxide below the 1 ppm level. J. Chromatogr. *91*, 539 (1974)

110. Tesch, J. W., Rehg, W. R., Sievers, R. E.: Microdetermination of nitrates and nitrites in saliva, blood, water and suspended particulates in air by gas-chromatography. J. Chromatogr. *126*, 743 (1976)

111. Uden, P. C., Henderson, D. E.: Non-fluorinated tetradentate β-keto-amines as derivatizing ligand for the gas-chromatographic analysis of Cu(II), Ni(II) and Pd(II). J. Chromatogr. *99*, 309 (1974)

112. Uden, P. C., Henderson, D. E., Burgett, C. A.: Trace copper and nickel chelate detection by electron capture gas-chromatography. Anal. Lett. *7*, 807 (1974)

113. Valentour, J. C., Aggarwal, V., Sunshine, I.: Sensitive gas-chromatographic determination of cyanide. Anal. Chem. *46*, 924 (1974)

114. Wells, J., Cimbura, G.: The determination of elevated bromide levels in blood by gas-chromatography. J. Forensic, Sci. *18*, 437 (1973)

115. Wolf, W. R.: Coupled gas-chromatography-atomic absorption spectrometry for the nanogram determination of chromium. Anal. Chem. *48*, 1717 (1976)

116. Young, J. W., Christian, G. D.: Gas-chromatographic determination of selenium. Anal. Chim. Acta *65*, 127 (1973)

117. Zaroegar, P., Mushak, P.: Quantitative measurements of inorganic mercury and organomercurials in water and biological media by gas-liquid chromatography. Anal. Chim. Acta *69*, 389 (1974)

118. (R) Walton, H. F.: Ion exchange and liquid column-chromatography. Anal. Chem. *50*, 36R (particular 37R–40R) (1978); Anal. Chem. *48*, 52R (particular 52, 53, 56R) (1976)

119. Adcock, L. H., Hope, W. G.: A method for the determination of tin in the range of 0,2–1,6 μg and its application to the determination of organotin stabilizer in certain foodstuff. Analyst *95*, 868 (1970)

120. Araki, S., Suzuki, S., Yamada, M.: Automatic ion-exchange chromatograph for analysis of alkali-metal and alkanline-earth metal mixtures, Talanta *19*, 577 (1972)

121. Araki, S., Suzuki, S., Hobo, T., Yamada, M.: Automated ion-exchange chromatography for the ultra-micro determination of alkali and alkaline-earth metals. Use of a double-jet hydrogen flame ionisation detector. Bunseki Kagaku *19*, 493 (1970); Anal. Abstr. *21*, 3235 (1971)

122. Brunfelt, A. O., Steinnes, E.: Determination of lutetium, ytterbium and terbium in rocks by neutron activation and mixed solvent anion-exchange chromatography. Analyst *94*, 979 (1969)

123. Cassidy, R. M.: A selective method for elemental sulfur analysis by high-speed liquid-chromatography. J. Chromatogr. *117*, 71 (1976)

124. Chao, T. T., Fishman, M. J., Ball, J. W.: Determination of traces of silver in water by anion exchange and atomic absorption chromatography. Anal. Chim. Acta *47*, 189 (1969)

125. Choe, D. S.: Determination of zirconium and hafnium by reverse-phase-chromatography, 2. Determination of zirconium and hafnium by using acetophenone as stationary phase. Punsok Hevalak *12*, 84 (1974); C. A. *82*, 164 400 z (1975)

126. Davenport, R. J., Johnson, D. C.: Determination of nitrate and nitrite by forced-flow liquid-chromatography with electrochemical detection. Anal. Chem. *46*, 1971 (1974)

127. Fisel, S., Franchevici, H., Balan, G.: Quantitative determination of tin by the chromatographic method on paper and on a cellulose column. Rev. Chim. Acad. Rep. Populaire Roumaine *6*, 175 (1961); Ref. Anal. Abstr. *9*, 609 (1962)

128. Fomia, M.: Rapid polarography in flowing solutions. Ann. Chim. *63*, 763 (1973); C. A. *83*, 21 591 x (1973)

129. Franks, M. C., Pullen, D. L.: Technique for the determination of trace anions by the combination of a potentiometric sensor and liquid-chromatography with particular reference to the determination of halides. Analyst *99*, 503 (1974)

130. Fritz, J. S., Goodkin, L.: Separation and determination of tin by liquid-solid-chromatography. Anal. Chem. *46*, 959 (1974)

131. Gaetani, E., Laureri, C. F., Mangia, A., Parolari, G.: High-pressure liquid-liquid partition chromatography of metal chelates of tetradentate β-ketoamines. Anal. Chem. *48*, 1725 (1976)

132. Hartkopf, A., Delumyea, R.: Use of the luminol reaction for metal ion detection in liquid-chromatography. Anal. Lett. 7, 79 (1974)

133. Hayden, J. A.: Determination of uranium in electro-refined plutonium by a combined ion-exchange and X-ray fluorescence technique. Talanta 14, 721 (1967)

134. Hani, S. M.: Chromatographic separation and spectrophotometric determination of uranium. Nucl. Sci. Abstr. 24, 38710 (1970); C. A. 74, 119 766x (1971)

135. Ivanova, I. M., Zorov, N. B.: Determination of indium impurity in metallic gallium. Vestn. Mosk. Univ. Khim. 15, 475 (1974); C. A. 82, 38 180q (1975)

136. Jones, D. V., Manahan, S. E.: Detection limits for flame spectrophotometric monitoring of high speed liquid-chromatographic effluents. Anal. Chem. 48, 1897 (1976)

137. Joshi, B. D., Patel, B. M.: Chemical separation and spectrographic determination of trace amounts of rare-earth elements in uranium. India At. Energy Comm. Bhabka. At. Res. Cent.(Rep.)) BARC 517 (1970); C. A. 75, 104748 (1971)

138. Karajannis, S., Ortner, H. M., Spitzy, H.: Column-chromatographic method for removal of the molybdenum matrix in determination of alkali and alkaline-earth metals in molybdenum and its compounds. Talanta 19, 903 (1972)

139. Kawabuchi, K., Kuroda, R.: A combined ion-exchange spectrophotometric method for the determination of molybdenum and tungsten in sea water. Anal. Chim. Acta 46, 23 (1969)

140. Kawazu, K.: Comparison of efficiency of cation-exchange resins in the chromatographic separation of metal ions with aqueous aceton-hydrochloric acid solutions. J. Chromatogr. 137, 381 (1977)

141. Kawazu, K., Fritz, J. S.: Rapid and continuous determination of metal ions by cation-exchange-chromatography. J. Chromatogr. 77, 397 (1973)

142. Korkisch, J., Sorio, A.: Determination of seven trace elements in natural waters after separation by solvent extraction and anion-exchange-chromatography. Anal. Chim. Acta 79, 207 (1975)

143. Korkisch, J., Sorio, A.: The determination of beryllium in geological and industrial materials by atomi-absorption-spectrometry after cation-exchange separation. Anal. Chim. Acta 82, 311 (1976)

144. Korkisch, J., Hübner, H., Steffen, I., Arrhenius, G., Fisk, M., Frazer, J.: Chemical analysis of manganes modules I. Determination of seven main and trace constituents after anion-exchange separation. Anal. Chim. Acta 83, 83 (1976)

145. Krainer, H., Ortner, H. M., Mueller, K., Spitzy, H.: Alkali and alkaline earth trace analysis of tungsten and tungsten compounds after tungsten matric separation of DEAE-Sephadex. Talanta 21, 933 (1974)

146. Kurbatova, V. I., Stepin, V. V., Ponosov, V. I., Novikova, E. V., Emasheva, G. N., Kalashnikova, L. Y.: Use of ion-exchange chromatography for the determining trace impurities in heat resistance alloys and ferroalloys. Zavod. Lab. 37, 413 (1971); C. A. 74, 150 696y (1971)

147. Lorber, K., Mueller, K., Spitzy, H.: Column extraction chromatography with dithizone in o-dichlorobenzene. Possibilities for separating metal traces in the nanogram region. Mikrochim. Acta 603 (1975)

148. Mangia, A., Parolari, G., Gaetani, E., Laureri, C. F.: High-pressure liquid-chromatography of dibenzo-18-crown-6 complexes with mercury (II) halides. Anal. Chim. Acta 92, 111 (1977)

149. Maura, G., Rinaldi, G.: Detection of traces of transition metals by peroxy compounds and sorption on a chelating resin. Anal. Chim. Acta 53, 466 (1971)

150. Matsui, H.: An automated method for the determination of trace amounts of metal ions by ion-exchange-chromatography. Determination of zinc(II) in waters. Anal. Chim. Acta 66, 143 (1973)

151. Matthews, A. D., Riley, J. P.: The determination of thallium in silicate rocks, marine sediments and sea water. Anal. Chim. Acta 48, (1969)

152. Mazzucotelli, A., Frache, R., Dadone, A., Baffi, F.: Ion-exchange separation and atomic-absorption determination of fifteen major, minor and trace elements in silicates. Talanta 23, 879 (1976)

153. McBride, L., Chorney, W., Skok, J.: Determination of boron in water, nutrient media, and Chlorella cells. Bot. Gaz. (Chicago) *133*, 103 (1972); C. A. *77*, 149 276 s (1972)

154. Neary, M. P., Seitz, P., Hercules, D. M.: A chemiluminescence detector for transition metals separated by ion exchange. Anal. Lett. *7*, 583 (1974)

155. Neirinckx, R., Adams, F., Hoste, J.: Determination of impurities in titanium dioxide by neutron activation analysis, Part III Determination of vanadium and aluminium in titanium and titanic by preseparation. Anal. Chim. Acta *47*, 173 (1969)

156. Neirinckx, R., Adams, F., Hoste, J.: Determination of impurities in titanium and titanium dioxide by neutron activation analysis, Part IV Determination of trace impurities in titanium dioxide single crystals. Anal. Chim. Acta *48*, 1 (1969)

157. Okubo, T.: Determination of rhenium by a combination of reversed-phase-chromatography and ultraviolet spectrophotometry. Bunseki Kagaku (Jap. Anal.) *24*, 600 (1975); Anal. Abstr. *30*, 4 B 137 (1976)

158. Paderina, F. P., Ol'shanova, K. M., Morozova, M. N.: Concentration of chromium and lead from highly dilute solutions by precipitation-chromatography. Nor. Metody Khim. Anal. Mater. No. 2, 63 (1971); C. A. *77*, 13 634z (1972)

159. Pagliai, V., Pozzi, F.: Determination of the metallic impurities in highly enriched uranium solutions by atomic absorption, with the preliminary separation of uranium by partition chromatography. Com. Naz. Energ. Nucl. RT/CHI (72), 10 (1972); C. A. *78*, 105 660d (1973)

160. Schwedt, G.: Application of high-pressure liquid chromatography in inorganic analysis, II. Separation of metal diethyldithiocarbamates by reversed-phase HPLC. Chromatographia *11*, 145 (1977)

161. Schwedt, G.: Inorganic trace analysis with high-pressure liquid -chromatography by the example of selenium. Z. Anal. Chem. *288*, 50 (1977)

162. Schwedt, G., Schwarz, A.: Application of high-pressure liquid-chromatography in inorganic analysis, III. Determination of selenium in drinking, natural and waste water. J. Chromatogr. *160*, 309 (1978)

163. Sen Gupta, J. G.: Determination of microgram amounts of the six platinum-group metals in iron and stary meteorites. Anal. Chim. Acta *42*, 481 (1968)

164. Seymour, M. D., Fritz, J. S.: Determination of metals in mixed hydrochloric and perchloric acids by forced-flow anion exchange-chromatography. Anal. Chem. *45*, 1394 (1973)

165. Seymour, M. D., Fritz, J. S.: Rapid, selective method for lead by forced-flow liquid-chromatography. Anal. Chem. *45*, 1632 (1973)

166. Seymour, M. D., Sickafoose, J. P., Fritz, J. S.: Application of forced-flow liquid chromatography to the determination of iron. Anal. Chem. *43*, 1734 (1971)

167. Shigematsu, T., Nishikaqa, Y., Hiraki, K., Goda, S., Tsujimoto, Y.: Fluorometric determination of cerium in sea-water. Bunseki Kagaku *20*, 575 (1971); Anal. Abstr. *23*, 2862 (1972)

168. Sixta, V., Miksovsky, M., Sulcek, Z.: Determination of barium in waters by atomic absorption spectrometry after ion-exchange-chromatography. Coellct. Czech. Chem. Commun. *38*, 3418 (1973)

169. Slezko, N. I., Chashchina, O. V., Synkova, A. G.: Chromatographic-determination of copper, cadmium, and lead in highly pure bismuth. Zavod. Lab. *41*, 13 (1975); C. A. *83*, 21 549q (1975)

170. Small, H., Stevens, T. S., Bauman, W. C.: Novel ion-exchange-chromatographic-method using conductometric detection. Anal. Chem. *47*, 1801 (1975)

171. Smith, J. D.: Spectrophotometric determination of traces of tin in rocks, sediments and soils. Anal. Chim. Acta *57*, 371 (1971)

172. Souliotis, A. G.: Simultaneous routine determination of copper and zinc in plants by neutron-activation analysis. Analyst *94*, 359 (1969)

173. Strelow, F. W. E.: Application of ion exchange-chromatography to accurate determination of lead, uranium, and thorium in tantaloniobates. Anal. Chem. *39*, 1454 (1967)

174. Sulcek, Z., Sixta, V.: The use of silica gel for the separation of traces of uranium. Anal. Chim. Acta *53*, 335 (1971)

G. Schwedt

175. Sulcek, Z., Sixta, V.: Separation of traces of bismuth from excess of iron, antimony, and copper. Collect. Czech. Chem. Commun. *37*, 1993 (1972)
176. Sulcek, Z., Pavondra, P., Kratchovil, V.: Analysis of metals and inorganic raw materials XVIII. Determination of bismuth in galena and pure lead. Collect. Czech. Chem. Commun. *34*, 3711 (1969)
177. Strelow, F. W. E., Liebenberg, C. J., Toerien, F. V. S.: Accurate silicate analysis based on separation by ion-exchange-chromatography. Anal. Chim. Acta *47*, 251 (1969)
178. Strelow, F. W. E., Liebenberg, C. J., Victor, A. H.: Accurate determination of the major and minor elements in silicate rocks based on separation by cation-exchange-chromatography on a single column. Anal. Chem. *46*, 1409 (1974)
179. Taylor, L. R., Johnson, D. C.: Determination of antimony using forced-flow liquid-chromatography with a coulometric detector. Anal. Chem. *46*, 262 (1974)
180. Uchida, T., Nagase, M., Kojimo, I., Iida, C.: A simple decomposition chelating resin separation for the determination of heavy metals in silicates by atomic absorption spectrometry. Anal. Chim. Acta *94*, 275 (1977)
181. Uden, P. C., Walters, F. H.: The high-pressure liquid-chromatographic separation of copper(II) and nickel(II) schiff base chelates on microparticulate silica. Anal. Chim. Acta *79*, 175 (1975)
182. Uden, P. C., Parees, D. M., Walters, F. H.: Analytical separations of schiff base chelates by reverse phase HPLC on a 10 micron C-18 bonded silica substrate. Anal. Lett. *8*, 795 (1975)
183. Uden, P. C., Bigley, I. E.: High-pressure liquid-chromatography of metal diethyldithiocarbamates with U. V. and D. C. argon-plasma emission spectroscopic detection. Anal. Chim. Acta *94*, 29 (1977)
184. Vens, M. D., Lauwerys, R.: Simultaneous determination of lead and cadmium in blood and urine by coupling ion-exchange resin-chromatography and atomic absorption spectrophotometry. Arch. Meh. Prof. Mech. Trav. Secur. Soc. *33*, 97 (1972); C. A. *77*, 70 973h (1972)
185. Wehner, H., Al-Murab, S., Stoppler, M.: Extraction-chromatographic separation of neptunium-239 from fission and activation products in the determination of microgram and submicrogram quantities of uranium. Radiochem. Radioanal. Lett. *13*, (1973)
186. Wheeler, G. L., Lott, P. F.: Rapid determination of trace amounts on selenium(IV), nitrite and nitrate by high pressure liquid-chromatography using naphthalene-2,3-diamine. Microchem. J. *19*, 390 (1974)
187. Willis, R. B., Fritz, J. S.: Determination of bismuth by forced-flow liquid-chromatography. Talanta *21*, 347 (1974)
188. Wolkoff, A. W., Larose, R. H.: Separation and detection of low-concentration of polythionates by high speed anion exchange liquid-chromatography. Anal. Chem. *47*, 1003 (1975)
189. (R) Brinkman, U. A. Th., de Vries, G., Kuroda, R.: Thin-layer-chromatographic data for inorganic substances. J. Chromatogr. *85*, 187 (1973); review 1938–1972
190. (R) Lederer, M.: Inorganic thin-layer-chromatography. Chromatogr. Rev. *9*, 115 (1967)
191. (R) Lesigang-Buchtela, M.: Thin layer-chromatography of inorganic ions. A review of the literature. Oesterr. Chemiker Ztg. *67*, 115 (1966)
192. Senf, H. J.: Applications of thin-layer-chromatography in inorganic analysis. Z. Chem. *6*, 102 (1966)
193. (R) Zweig, G., Sherma, J.: Paper and thin-layer-chromatography. Anal. Chem. *50*, No. 5, 50R, resp. 60R (1978); Anal. Chem. *48*, No. 5, 66R, resp. 76R (1976)
194. Abe, S., Weisz, H.: The rapid separation and determination of uranium(IV) by use of DEAE anion-exchange paper and the ring-oven technique. Mikrochim. Acta 550 (1970)
195. Agrinier, H.: Application of paper-chromatography for the determination of trace elements in geological materials. Dosage Elem. Etat Traces Roches Autres Subst. Miner. Nat. Actes Colloq. 1968, 309 (1970); Ref. C. A. *75*, 104 750j (1971)
196. Aleskovskaya, V. N., Aleskovskii, V. B.: Determination of microamounts of certain cations by peak precipitation-chromatography on paper impregnated with slightly soluble diethyldithiocarbamates. Zh. Anal. Khim. *25*, 243 (1970); Ref. C. A. *73*, 10 423m (1970)
197. Andreasen, B., Bohlbro, M., Crossland, H., Dennak, V.: Slip-chromatography: a semiquantitative TLC technique. J. Chem. Educ. *53*, 772 (1976); C. A. *86*, 42 508c (1977)

198. Bäumler, J., Rippstein, S.: Micro-determination of mercury. Mitt. Geb. Lebensm.-Hyg. *54*, 472 (1963); Ref. Anal. Abstr. *12*, 1099 (1965)
199. Baranowski, R., Kot, B., Baranowski, I., Gregorowicz, Zb.: 2,2'-Diquinoxalyl as a new reagent for determination of bivalent tin in thin-layer-chromatography. Microchem. J. *20*, 1 (1975)
200. Barna, R. K., Baishya, N. K.: Spectrophotometric determination of lead, copper, and bismuth in a mixture after separation by paper-chromatography, Current Sci. (India) *37*, 434 (1968); Ref. C. A. 92 657v (1968)
201. Baudot, P., Monal, J. L., Livertonx, M. H., Truhant, R.: Identification of toxic metals after extraction and thin-layer-chromatography of their dithizonates. Toxicological applications. J. Chromatogr. *128*, 141 (1976)
202. Becherer, K.: Quantitative paper-chromatographic determination of alkali metals. Oesterr. Akad. Wiss. Math. Naturw. Kl., Sitzber. Act. I 175, No. 4–6, 107 (1966); Ref. C. A. *69*, 15 684z (1968)
203. Ben-Dor, L., Jungreis, E.: Ultramicro spectrofluorimetric determination of aluminium. Isr. J. Chem. *8*, 951 (1970); Ref. C. A. *75*, 58 338f (1971)
204. Beszterda, A.: Paper-chromatographic determination of microgram amounts of sulphates. Chem. Anal. (Warsaw) *14*, 341 (1969); Ref. Anal. Abstr. *19*, 195 (1970)
205. Bhatnagar, R. P., Bhattacharya, K. N.: Quantitation of uranyl ion by paper-chromatographic technique. J. Indian Chem. Soc. *53*, 931 (1976); Ref. C. A. *87*, 62 066f (1977)
206. Biala, D.: Determination of lead in the blood by thin-layer-chromatography without mineralization. Brometol. Chem. Toksykol. *7*, 63 (1974); Ref. C. A. *81*, 58 838t (1974)
207. Billinski, S., Klimek, J., Misiuna, D.: Methodical investigations on the usefulness of 2-hydrazinothiazole derivatives for detecting microgram amounts of metal ions on paper-chromatography, II. Determination of Mn^{2+}, Co^{2+}, Ni^{2+} ions by direct planimetry of spots. Ann. Univ. Mariae Curie-Sklodowska Sect. D *25*, 415 (1970); Ref. C. A. *76*, 148 452p (1972)
208. Boenig, G.: Determination of total strontium in vegetable material and sea water by circular paper-chromatography. Omagiu Raluca Ripan 157 (1966); C. A. *68*, 18 410x (1968)
209. Boenig, G., Heigener, H.: Paper-chromatographic separation of lead in vegetables, its identification and determination. Landwirtsch. Forsch. *19*, 117 (1966); Ref. Z. Anal. Chem. *231*, 457 (1967)
210. Boenig, G., Heigener, H.: Microdetermination of vanadium in biological substances by selective paper-chromatography. Landwirt. Forsch. *25*, No. 2, 139 (1972); C. A. *77*, 722 216n (1972)
211. Boenig, G., Heigener, H.: Paper-chromatographic analysis of thallium in organic material. Landwirt. Forsch. *26*, 81 (1973); C. A. *79*, 28 002g (1973)
212. Bogdanova, E. G., Aleskovskaja, V. N.: Peak-chromatographic determination of trace amounts of platinum group metals. Zh. Anal. Khim. *29*, 1857 (1954); C. A. *82*, 67 778w (1975)
213. Boichinova, E. S., Aleskovskii, V. B.: Determination of trace elements by the peak heights on paper-chromatograms. I. Determination of copper and nickel ions. Tr. Leningr. Tekhnol. Inst. im. Lensoveta *48*, 94 (1958); C. A. *59*, 9 308e (1963)
214. Brunstad, J. W.: Colorimetric determination and thin layer identification of boron as boronic acid in caviar. J. Assoc. Offic. Anal. Chemists *51*, 987 (1968)
215. Cernikova, M., Konrad, B.: Microestimation of zinc in human blood serum. Biochem. Biophys. Acta *71*, 190 (1963)
216. Chikui, S.: Electrometric determination in chromatographic development, III. Determination of alkaline earth metals during ascending development on paper, and their separation from alkali metal ions. Jap. Analyst *20*, 167 (1971); Anal. Abstr. *23*, 93 (1972)
217. Chiotis, E. L., Welford, G. A., Morse, R. S.: Trace determination of nickel by combining paper-chromatography with radiometric precipitation. Mikrochim. Acta 297 (1968)
218. Chistyakov, N. M., Blagoveshchenskya, Z. I.: Determination of copper, cobalt and nickel in water by paper-chromatography. Gigiena i Sanit. *28*, 58 (1963); C. A. *59*, 7 238g (1963)
219. Connolly, J. F., Maguire, M. F.: An improved chromatographic method for determining trace elements in foodstuffs. Analyst *88*, 125 (1963)

220. Covello, M., Ciampa, G.: Quantitative analysis of lithium, potassium and caesium chloride by paper-chromatography and flame spectrophotometry. J. Chromatogr. *20*, 201 (1965)

221. Covello, M., Ciampa, G., Manne, F.: Separation and spectrophotometric determination of alkali metals in mineral water. Rend. Accad. Sci. Fis. Mat. (Soc. Nazl. Sci. Napoli) *33*, 325 (1966); C. A. *69*, 5 109b (1968)

222. Cremer, E., Seidl, E.: Separation of radioactive ánions by thin-film sorptography in the range below 10^{-14} g; Chromatographia *3*, 17 (1970)

223. Draignand, M.: Application of paper-chromatography to the determination of trace elements in silicates. Bull. Soc. Fr. Ceram. *83*, 27 (1969); C. A. *72*, 24 100v (1970)

224. Dutkiewicz, T., Podoski, A., Komstaszumska, E., Liszczyna, I.: Chromatographic separation and colorimetric determination of mercury, cadmium, cobalt, nickel, and zinc as dithizonates. Bromatol. Chem. Toksykol. *6*, 381 (1973); C. A. *80*, 127 770h (1974)

225. Fedorova, N. E., Kovalenko, P. N.: Determination of small amounts of niobium by partition paper-chromatography in ores and minerals. Sovrem. Metody Khim. Tekhnol. Kontr. Prvizood. 23 (1968); C. A. *72*, 8 960v (1970)

226. Fedorova, N. E., Stepanova, N. P., Eiger, V. I.: Paper-chromatographic determination of trace amounts of zirconium and hafnium. Zavod. Lab. *39*, 411 (1973); C. A. *79*, 38 263g (1973)

227. Floret, G., Massa, V.: Determination of cobalt, copper, nickel and manganese after chromatographic separation of their complexes with 1-(2-pyridylazo)-2-naphthol. Trav. Soc. Pharm. Montpellier *28*, 129 (1969); C. A. *70*, 31.731p (1969)

228. Fomin, A. A.: Determination of trace elements in biological materials by paper partition-chromatography. Mikroelement. Biosfere Ikh Prismen, S-kh. Med. Sib. Dal'nejo Wastoka, D Dokl. Sib. Konf. 4th 1972 (Publ. 1973) 490; C. A. *83*, 4 045e (1975)

229. Frei, R. W.: A thin-layer-chromatographic method for determination of trace elements in cereal. J. Chromatogr. *34*, 563 (1968)

230. Frei, R. W., Miketakova, V.: A reflectance spectroscopic study of 4-(2-thiazolyl-azo)resorcinol (TAR) as a spray reagent for cobalt, copper and nickel on thin-layer chromatograms. Mikrochim. Acta 290 (1971)

231. Frei, R. W., Ryan, D. E.: Trace metal analysis by combined thin-layer chromatography and reflectance spectroscopy. Anal. Chim. Acta *37*, 187 (1967)

232. Frei, R. W., Stillman, H.: A combination of thin-layer chromatography and emission spectroscopy for the determination of alkaline earth metals. Mikrochim. Acta 184 (1970)

233. Frei, R. W., Stockton, C. A.: A combination of ring-oven and circular-chromatography for trace metal analysis. Mikrochim. Acta 1196 (1969)

234. Frei, R. W., Liiva, R., Ryan, D. E.: Reflectance spectroscopic determination of cobalt, nickel, and copper with pyridine-2-aldehyde-2-quinolylhydrazone. Can. J. Chem. *46*, 167 (1968)

235. Frodyma, M. M., Zaye, D. F., Van Lien, T.: Cation analysis by thin-layer chromatography and reflectance spectroscopy, Part II. The determination of copper, nickel and zinc. Anal. Chim. Acta *40*, 451 (1968)

236. Gaibakyan, D. S., Egikyan, R. T.: Thin-layer chromatography of rare elements IX. Identification and determination of gold III, selenium IV and tellurium IV in an alkaline medium. Ann. Khim. Zh. *23*, 16 (1970); C. A. *73*, 31 258b (1970)

237. Gauchev, N., Gaucheca, A.: Microchemical analysis on paper, microquantitative determination of Fe. Compt. Rend. Acad. Bulgare Sci. *18*, 441 (1965); C. A. *63*, 9 050a (1965)

238. Gel'man, E. M.: Determination of small amounts of selenium and tellurium in mineral materials using paper-chromatographic separation. Zh. Anal. Khim. *23*, 736 (1968); C. A. *69*, 48 993c (1968)

239. Gniazdowski, M., Wasiak, T., Filipowicz, B.: Direct colorimetric microdetermination of phosphates in chromatographic spots. Chem. Anal. (Warsaw) *16*, 1367 (1971); Anal. Abstr. *23*, 1390 (1972)

240. Gorbach, G.: Analysis of metal traces by means of microchromatography. Metal Catal. Liquid Oxid. SIK Symp. 1967 (publ. 1968), 67; C. A. *71*, 45 369j (1969)

241. Graham, R. J. T., Bark, L. S., Tinsley, D. A.: Quantitative thin-layer chromatography on liquid anion exchangers Part I. An investigation into some of the parameters involved in the direct densitometric determination of zinc. J. Chromatogr. 39, 211 (1969)

242. Graham, R. J. T., Bark, L. S., Tinsley, D. A.: Quantitative thin-layer chromatography on liquid anion exchangers, Part II. A comparison of a spot removal method with a *in situ* direct densitometric method. J. Chromatogr. 39, 218 (1969)

243. Grassini, G., Alberti, G.: A very sensitive fluorescence test for uranium and its use in the paper-chromatographic estimation of it in natural water. Microchem. J., Symp. Ser. 2, 285 (1962); C. A. 58, 6 573 e (1963)

244. Gregorowicz, Z., Kulicka, J., Suwinska, T.: Separation and detection of trace amounts of some heavy metal dithizonates by thin-layer chromatography. Chem. Anal. (Warsaw) 16, 169 (1971); Anal. Abstr. 21, 3212 (1971)

245. Handa, A. C., Johri, K. N.: Micro determination of sulfide, sulfite, sulfate and thiosulfate by thin-layer chromatography and ring-colorimetry. Talanta 20, 219 (1973)

246. Hara, N.: Simple determination of lead in air using thin-layer chromatographic plates. Ind. Health 11, 155 (1973); C. A. 81, 29 085q (1974)

247. Hara, N., Matsumura, Y.: Simple ion-exchange thin-layer chromatographic technique for the determination of traces of metals: Kogyo Kagaku Zasshi 74, 364 (1971); C. A. 74, 150 838w (1971)

248. Hara, N., Matsumura, Y.: Simple ion-exchange thin-layer chromatographic technique for the determination of traces of metals. Ind. Health 9, 72 (1971); C. A. 78, 33 493b (1973)

249. Hejtmanek, M.: Quantitative chromatography of inorganic substances I. Principles of photometric evaluation of paper-chromatography. Sb. Vysoke Skoly Chem-Technol. v Praze, Oddil Fak. Anorg. Org. Technol. 4, 63 (1960); C. A. 61, 1 246h (1964)

250. Hejtmanek, M.: Quantitative chromatography of inorganic substances, II. Chromatographic microdetermination of nickel, cobalt and copper by photometry *in situ*. Sb. Vysoke Skoly Chem.-Technol. v Praze, Oddil Fak. Anorg. Org. Technol. 4, 69 (1960); C. A. 61, 1 247 a (1964)

251. Hejtmanek, M., Hozmanova, E.: Spectrophotometric determination of caesium and rubidium as picrates after paper-chromatographic separation. Mikrochim. Acta 97 (1966)

252. Hohmann, E., Rafizadeh, M., Specker, H.: Chromatographic separations of the rare earths. Z. Anal. Chem. 286, 50 (1977)

253. Hui, K. S., Davis, B. A., Boulton, A. A.: The separation by thin-layer chromatography of trace metals as their tetraphenylporphyrin chelates. J. Chromatogr. 115, 581 (1975)

254. Johri, K. N., Mehra, H. C.: Trace metal analysis by combined thin-layer chromatography incorporating fluorescent support and ring-oven colorimetry. Microchem. J. 15, 642 (1970)

255. Johri, K. N., Mehra, H. C.: Inorganic analysis by combinded thin-layer chromatography and ring-oven technique incorporating fluorescent support. Separ. Sci. 6, 741 (1971)

256. Johri, K. N., Mehra, H. C.: Microdetermination of gold (III), silver (I) and ruthenium (III) by ring colorimetry after separation by thin-layer chromatography. Mikrochim. Acta 807 (1970)

257. Johri, K. N., Kaushik, N. K., Bakshi, K.: Thin-layer chromatographic separation of copper (II), nickel (II), and cobalt (II) as thio-carbonato-complexes and determination by ring colorimetry. Chromatographia 5, 326 (1972)

258. Johri, K. N., Mehra, H. C., Kaushik, N. K.: Determination of GeIV, SnII, PbII and ZnII, CdII, HgII by ring colorimetry after separation by thin-layer chromatography. Chromatographia 3, 347 (1970)

259. Jung, K., Specker, H.: Chromatographic separation and detection of the rare earth elements in ores, technical products and pure elements. Z. Anal. Chem. 288, 28 (1977)

260. Kamin, G. J., O'Laughlin, J. W., Banks, C. V.: Separation and determination of zirconium in niobium using methylenebis(di-n-hexylphosphine oxide). J. Chromatogr. 31, 292 (1967)

261. Kaukare, P., Suovaniemi, O.: A simple method for determination of phosphate from thin-layer chromatographic plates. J. Chromatogr. 62, 485 (1971)

262. Kawagaki, K., Kadoki, H., Ono, M.: Separation and determination of indium by paper-chromatography. Nippon Kagaku Zasshi 90, 1282 (1969); C. A. 72, 74 421w (1970)

263. Kielczewski, W., Supinski, J.: Determination of microgram amounts of cobalt by paper impregnation method. Chem. Anal. (Warsaw) 8, 59 (1963); Anal. Abstr. 10, 5174 (1963)

264. Kielczewski, W., Matusiewicz, K.: Determination of microgram amounts of magnesium by impregnation method on a filter paper. Chem. Anal. (Warsaw) 13, 787 (1968); Anal. Abstr. 17, 3332 (1969)

265. Kokk, K. Y., Aleskovskii, V. B.: Determination of copper in micro amounts of cadmium selenide by paper precipitation-chromatography. Izo. Vysshikh. Uchebn. Zavedenii, Khim. i Khim. Teknol. 7, 564 (1964); C. A. 62, 3 392f (1965)

266. Kostikov, A. P., Egorova, S. N., Bulenkov, T. I.: Separation and determination of copper, nickel and cobalt by thin-layer chromatography combined with reflectance spectrophotometry. Izo. Vysshik. Uchebn. Zaved. Khim. Khim. Tekhnol. 17, 1257 (1974); C. A. 82, 38 218h (1975)

267. Krzeczkowska, I., Bilinski, S., Klimek, J., Misiuna, D.: 2-Hydrazinothiazole derivatives for the detection of μg amounts of Cu^{2+}, Co^{2+}, Fe^{2+} and Fe^{3+} by paper-chromatography, II. Quantitative determination of Cu^{2+} and Co^{2+} by direct planimetry of spots. Ann. Univ. Mariae Curie-Sklodowska Sect. D 20, 123 (1965), C. A. 67, 96 473p (1967)

268. Kuz'micheva, M. N.: Separate determination of chromium and nickel in the air by paper-chromatography. Gig. Sanit. 36, 70 (1971) C. A. 76, 37 091a (1972)

269. Lien, V. T., Handy, C. A.: In situ fluorometric determination of alkaline earth metal ions resolved on paper. Anal. Lett. 7, 267 (1974)

270. Mah, D. C., Mah, S., Tupper, W. M.: Circular paper-chromatographic method for determining trace amounts of cobalt, copper, nickel, and zinc in rocks and soils. Can. J. Earth Sci. 2, 33 (1965); C. A. 63, 6 304b (1965)

271. Maly, E.: Determination of the total silica level in ashed biological material by paper-chromatography. Pracooni Lekar 18, 220 (1966); C. A. 65, 12 542h (1966)

272. Markina, N. A.: Chromatographic separation of trace amounts of copper, nickel and cobalt in presence of iron in the air. Gig. Sanit. 36, 65 (1971); C. A. 76, 37 093c (1972)

273. Markina, N. A.: Determination of copper, cobalt and nickel directly on a chromatogram using a SF-10 spectrophotometer. Gig. Sanit. 7, 82 (1973); C. A. 79, 121 563a (1973)

274. Maslowska, J., Soloniewicz, R.: The use of thioacetamide for approximate determination of traces of heavy metals by paper-chromatography. Zeszyty Politech. Lodz Chem. 13, 27 (1963); C. A. 61, 4 938g (1964)

275. Massa, V.: Determination of mercury chromatography and photodensitometry of dithizonate. Control of mercuric oxide ointments: Trav. Soc. Pharm. Montpellier 28, 203 (1968); C. A. 70, 99 675d (1969)

276. Massa, V.: Determination of silver in the presence of other metals by chromatography and photodensitometry of its dithizonate derivatives. Application to drug control. Trav. Soc. Pharm. Montpellier 29, 221 (1969); C. A. 73, 69 930c (1970)

277. Massa, V.: Photodensitometric determination of lithium in the presence of other alkaline metals following thin-layer chromatography. Trav. Soc. Phram. Montpellier 29, 209 (1969); C. A. 72, 139 246q (1970)

278. Massa, V.: Microdetermination of manganese by chromatography and photodensitometry of its complex with 1-(2-pyridylazo)-2-naphthol. Trav. Soc. Pharm. Montpellier 29, 257 (1969); C. A. 73, 113 036s (1970)

279. Massa, V.: Determination of cadmium in the presence of zinc and other metals by dithizonate-chromatography and photodensitometry. Farmaco Ed. Prat. 25, 332 (1970); C. A. 73, 38 607y (1970)

280. Massa, V., Susplugas, P., Salabert, J.: Microdetermination of alkaline earth cations by fluorometry of their thin-layer chromatograms. Application to the analysis of vegetable ash. Trav. Soc. Pharm. Montpellier 34, 175 (1974); C. A. 81, 165 733b (1974)

281. Miketukova, V., Frei, R. W.: The thin-layer chromatographic properties of trace metals. I. Separation of some metals with special reference to lead. J. Chromatogr. 47, 427 (1970)

282. Mora, G. A.: Metallochromic indicators in inorganic paper-chromatography for the determination of trace elements in foods. Infom. Quim. Anal. (Madrid) 20, 91 (1966); C. A. 66, 27 755z (1967)

283. Murty, A. S. R.: Application of thiourea to the determination of selenium by paper-chromatography. J. Kamatak Univ. *17*, 19 (1972); C. A. *82*, 67 769u (1975)

284. Myasoedova, G. V., Volymets, M. P., Koreshinkova, T. A., Belyaev, Y. I., Dubrova, T. V.: Thin-layer chromatography of the noble metals 9. Use of modified cellulose for the determination of microgram amounts of silver. Zh. Anal. Khim. *29*, 2252 (1974); C. A. *82*, 179 895 p (1975)

285. Pasechnova, R. A., Mokhov, A. A., Yudin, B. F.: Determination of tellurium by Bismuthiol II in the presence of cadmium using paper peak-chromatography. Zh. Anal. Khim. *29*, 2292 (1974); C. A. *82*, 148 991 n (1975)

286. Paster, A., Kabacoff, B.: The quantitative analysis of free elemental sulphur by thin-layer chromatographic spectrodensitometry. J. Chromatogr. Sci. *14*, 572 (1976)

287. Pawlaczyk, J., Sierzant, M.: Thin-layer chromatography of some metallic contaminations in pharmaceutical preparations. Farm. Pol. *30*, 37 (1974); C. A. *81*, 82 473r (1974)

288. Pensionerova, V. M., Pankova, V. E.: Microchemical determination of selenium and tellurium in sulfide minerals using paper partition-chromatography. Byul. Nauchn.-Tekh. Inform. Min. Geol. SSSR, Ser. Izuch. Veshchestv. Sortova Miner. Syr'ya Tekhnol. Obogashch. Rud. *3*, 37 (1967); C. A. *69*, 56 791a (1968)

289. Plamondon, J.: Rapid determination of uranium in geochemical samples by paper-chromatography. Econ. Geol. *63*, 76 (1968); C. A. *69*, 15 918d (1968)

290. Popper, E., Florean, E., Marcu, P., Sosa, E.: New organic reagent in paper-chromatography. Application to inorganic substances. Determination of cadmium, bismuth, and lead. Rev. Roumanine Chim. *11*, 283 (1966); C. A. *65*, 1 353c (1966)

291. Puroshottam, D.: Chromatographic separation and estimation of traces of uranium in soils and plant ash, J. Sci. Ind. Res. (India) *19 B*, 449 (1960); C. A. *60*, 9 907h (1964)

292. Razina, I. S., Viktorova, M. E.: Determination of submicrogram amounts of osmium and ruthenium after separation by partition paper-chromatography. Metody Anal. Redkometal. Mineral. Rud. Gom. Rorod. *1*, 53 (1971); C. A. *80*, 33 582w (1974)

293. Razina, I. S., Viktorova, M. E.: Use of paper partition-chromatography for the determination of ruthenium and osmium in mineral raw materials. Zh. Anal. Khim. *29*, 2254 (1974); C. A. *82*, 179 914u (1975)

294. Reimann, K.: Quick identification of heavy metal ions in water and sludge by means of thin-layer chromatography using dithizone. Z. Wasser Abwasser Forsch. *5*, 3 (1972); C. A. *77*, 9 470a (1972)

295. Romanowski, H., Radzik, D.: Chromatographic separation of zinc with 8-mercaptoquinoline and its complexometric determination in toxicological analysis. Dissertationes Pharm. *17*, 559 (1965); C. A. *64*, 18 002a (1966)

296. Ryabchikov, D. I., Volynets, M. P., Kopneva, L. A.: Thin-layer chromatography in inorganic analysis, IV. Microchemical analysis of cosmic dust. Zh. Anal. Khim. *24*, 72 (1969); C. A. *70*, 92 895 s (1969)

297. Schwedt, G., Lippmann, Ch.: Rapid thin-layer chromatographic test for toxic metals in water. Dtsch. Lebensmittel-Rundschau *70*, 204 (1974)

298. Seiler, H.: Inorganic thin-layer chromatography, 7. Quantitative determination of thin-layer chromatographic separated cations. Helv. Chim. Acta XLVI, 2629 (1973)

299. Sheimina, R. J., Khalimova, U. Kh., Begmatova, M. P.: Determination of residual quantities of magnesium chlorate in cotton seeds, their pods and the cotton oil cake. Khim. Sel'sk. Khoz. *14*, 66 (1976); C. A. *85*, 138 116d (1976)

300. Shmanenkova, G. J., Romantseva, T. J., Pleshakova, G. P.: Determination of rare earths in cobalt- and copper-based magnetic alloys. Zh. Anal. Khim. *29*, 1545 (1974); C. A. *82*, 51 057z (1975)

301. Smoczkiewicz, A., Mizgalski, W.: Determination of zinc in blood serum. Bull. Soc. Annin. Sci. Lettres Pozman. Ser. C *10*, 61 (1960); C. A. *55*, 24 895c (1961)

302. Smoczkiewicz, A., Smoczkiewicz-Szczepanska, T.: Microdetermination of copper in pharmaceutical materials and preparations. Acta Polon. Pharm. *19*, 59 (1962); C. A. *57*, 4 763d (1962)

303. Süry, P.: Quantitative microanalysis of copper and aluminium by paper-chromatography. Z. Anal. Chem. *250*, 190 (1970)

304. Sulser, H.: Paper-chromatographic detection and approximate determination of trace metals in food dyes. Mitt. Gebiet Lebensm. Hyg. *57*, 66 (1966)

305. Suzuki, M., Kaiho, F., Takitani, S.: Thin-layer chromatography of inorganic compounds. XIV. Colorimetric determination of metal dithizonates by elution from the thin-layer plate. Eisei Kagaku (J. Hyg. Chem.) *21*, 47 (1975); C. A. *83*, 187 955u (1975)

306. Takitani, S., Fukuoka, N., Mitsuzawa, Y.: Inorganic thin-layer chromatography, VIII. Determination of metal ions by densitometry. Japan Analyst *15*, 840 (1966); Anal. Abstr. *15*, 2485 (1968)

307. Takitani, S., Suzuki, M., Namai, T., Nagata, F.: Thin-layer chromatography of inorganic compounds, XIII. Determination of metallic dithizonates by densitometry of thin-layer chromatograms. Eisei Kagaku *16*, 83 (1970); C. A. 72 748e (1970)

308. Tewari, S. N., Bhatt, N.: Separation and identification of metal dithizonates by thin-layer chromatography and its application in toxicological analysis. Mikrochim. Acta 337 (1973)

309. Thielemann, H.: Thin-layer chromatographic detection limits (semiquantitative determination) of toxic metal ions (lead, mercury, zinc, copper). Z. Chem. *15*, 110 (1975); see also Z. Anal. Chem. *275*, 206 (1975)

310. Thielemann, H.: Thin-layer chromatographic separation, identification and determination limits as well as the semiquantitative determination of halide ions on activated, commercially available films. Z. Chem. *16*, 283 (1976)

311. Todorova, T., Getcheva, T., Mandova, B.: Quantitative microdetermination of nickel on impregnated chromatographic-paper. Mikrochim. Acta 509 (1974)

312. Turina, N.: Rapid determination of lead in used lubricating oils on thin-layer chromatograms. J. Chromatogr. *93*, 211 (1974)

313. Turina, N.: Thin-layer chromatography of tin on a cellulose support using fluorescent chloro-complexes. Chromatographia *9*, 513–516 (1976)

314. Turina, N., Turina, S.: Trace analysis of lead in mineral oil by TLC. Chromatographia *10*, 97 (1977)

315. Vagina, N. S., Volymets, M. P.: Thin-layer chromatography in inorganic analysis. II. Separation and determination of rare earths and other impurities during the analysis of lanthanide preparations of various degree of purity. Zh. Anal. Khim. *23*, 521 (1968); C. A. *69*, 32 705z (1968)

316. Viktorova, M. E., Isaeva, K. G.: Separation of nickel, cobalt, copper and iron by paper-chromatography and their quantitative determination in sulphide ores and rocks. Zh. Anal. Khim. *25*, 1140 (1970); C. A. *73*, 105 109c (1970)

317. Volymets, M. P., Vagina, N. S., Fomina, T. V., Fakina, L. K.: Thin-layer chromatography in inorganic analysis, VI. Determination of rare earth elements in uranyl nitrate of high purity. Zh. Anal. Khim. *24*, 1477 (1969); C. A. *72*, 18 198s (1970)

318. Webb, R. A., Hallas, D. G., Stevens, H. M.: The determination of iron, manganese, zinc and copper in plant material by paper-chromatography and reflectance densitometry. Analyst *94*, 794 (1969)

319. Yamane, Y., Miyazaki, M., Iwase, H., Muramatsu, S.: Analysis of metals in water. I. Detection of metals by dithizone extraction and thin-layer chromatography. Eisei Kagaku *13*, 212 (1967); C. A. *68*, 43 060r (1968)

320. Yamane, Y., Miyazaki, M., Iwase, H.: Analysis of metals in water II. Analysis of aluminium, beryllium and chromium by oxine, 2-methyloxine extration and by thin-layer chromatography. Eisei Kagaku *14*, 106 (1968); C. A. 45 939 d (1968)

321. Yamane, Y., Miyazaki, M., Iwase, H., Takeuchi, T.: Analysis of metals in water, III. Total analysis of metals by dithizone, oxine, 2-methyloxine extraction and by thin-layer chromatography. Eisei Kagaku *16*, 254 (1970); C. A. *75*, 52 645g (1971)

322. Alimarin, I. P., Bolshova, T. A.: Separation and concentration of elements by reversed-phase partition-chromatography. Pure Appl. Chem. *31*, 493 (1972)

323. Alimarin, I. P., Bolshova, T. A., Ershova, N. I., Polinskaya, M. B.: The use of partition-chromatography for concentration traces of elements. Zh. Anal. Khim. *24*, 26 (1969); Anal. Abstr. *19*, 109 (1970)

324. Autokol'skaya, I. I., Myasoedova, G. V., Bol'shakova, I., Ezermitskaya, M. G., Volymets, M. P., Karyakin, A. V., Savvin, S. B.: Concentration and separation of elements on chelate sorbents. A sorbent for the noble metals based on the copolymer of styrene and 3(5)-methylpyrazole. Zh. Anal. Khim. *31*, 742 (1976); C. A. *85*, 86 640y (1976)

325. Bauman, A. J., Weetall, H. H., Weliky, N.: Coupled ligand-chromatography, applications to trace element collection and characterization. Anal. Chem. *39*, 932 (1967)

326. Braun, T., Farag, A. B.: Plasticized open-cell polyurethane foam as a universal matrix for organic reagents in trace element preconcentration I. Collection of silver traces on dithizone foam. Anal. Chim. Acta *69*, 85 (1974)

327. Braun, T., Farag, A. B.: Reversed-phase foam-chromatography. Chemical enrichment and separation of gold in the tributyl-phosphate-thiourea-perchloric acid system. Anal. Chim. Acta *65*, 115 (1973)

328. Brits, R. J. N., Smit, M. C. B.: Determination of uranium in natural water by preconcentration on anion-exchange resins and delayedneutron counting. Anal. Chem. *49*, 67 (1977)

329. Figura, P., McDuffle, B.: Characterization of the calcium form of Chelex-100 for trace metal studies. Anal. Chem. *49*, 1950 (1977)

330. Ghosh, A. K., Rajqar, D. P., Bandyopadhyay, P. K., Gosh, S. K.: Concentration technique for determination of air pollutants at sub-micro level. A new technique for concentration of carbon monoxide. J. Chromatogr. *117*, 29 (1976)

331. Going, J. E., Wosenberg, G., Andrejat, G.: Preconcentration of trace metal ions by combined complexation-anion-exchange I. Cobalt, zinc and cadmium with 2-(3'-sulfobenzoyl)-pyridine-2-pyridylhydrazone. Anal. Chim. Acta *81*, 349 (1976)

332. Guedes da Mota, M. M., Römer, F. G., Griepink, B.: Automated separation and preconcentration of copper II from natural waters using a column treatment. Z. Anal. Chem. *287*, 19 (1977)

333. Kawabuchi, K., Kanke, M., Muraoka, T., Yamauchi, M.: Ion-exchange concentration on a chelating resin and atomic absorption spectrophotometric determination of heavy metals in geochemical samples. Bunseki Kagaku *25*, 213 (1976); C. A. *86*, 100 299y (1977)

334. Lee, C., Kim, N. B., Lee, I. C., Chung, K. S.: The use of a chelating resin column for preconcentration of trace elements from sea-water in their determination by neutron-activation analysis. Talanta *24*, 241 (1977)

335. Lin, J. W., Janauer, G. E.: Selective separations by reactive ion exchange, III. Preconcentration and separation of oxo anions. Anal. Chim. Acta *79*, 219 (1975)

336. Meier, H., Ruckdeschel, A., Zimmerhackl, E., Albrecht, W., Bösche, D., Hecker, W., Menge, P., Unger, E., Zeitler, G.: For application of paper-chromatography for enrichment of trace elements in the geochemistry. Mikrochim. Acta 852 (1969)

337. Molnar, F.: Anion exchange concentration of rare-earth impurities in rare-earth materials for analytical purposes. Anal. Appl. Rare Earth Mater., NATO Adv. Study Inst. 1972 (Publ. 1973), 55

338. Nikitin, M. K., Maslov, V. A., Serkova, V. A.: Concentration of gold from solution by partition-chromatography. Sb. Tr. Nauch.-Issled. Proekt.-Knstr. Inst. No. 1, 19 (1971); C. A. *79*, 152 516s (1973)

339. Razina, I. S., Viktorova, M. E.: Concentration and separation of the platinum group elements and gold using ion-exchange and paper partition-chromatography. Zh. Anal. Khim. *25*, 1160 (1970); C. A. *73*, 115 979a (1970)

340. Savvin, S. B., Autokolskaja, I. I., Myasoedova, G. V., Bolshakova, L. I., Shvoeva, D. P.: Chelate sorbents for concentration and separation of noble metals. J. Chromatogr. *102*, 287 (1974)

341. Schiller, P., Cook, G. B.: Determination of trace amounts of gold in natural sweet waters by non-destructive activation analysis after preconcentration. Anal. Chim. Acta *54*, 364 (1971)

G. Schwedt

342. Schulek, E., Remport-Horvath, Z., Laszity, A., Koros, E.: Enrichment of traces of metals on carboxycellulose cation exchanger. Magy. Kem. Foly. *75*, 58 (1969); Anal. Abstr. *18*, 4470 (1970)
343. Schulek, E., Remport-Horvath, Z., Laszity, A., Koros, E.: Collection of traces of metals on carboxycellulose cation exchanger. Talanta *16*, 323 (1969)
344. Vanderborght, B. M., van Grieken, R. E.: Enrichment of trace metals in water by adsorption on activated carbon. Anal. Chem. *49*, 311 (1977)
345. Van Grieken, R. E., Bresseleers, C. M., Vanderbought, B. M.: Chelex-100 ion-exchange filter membranes for preconcentration in X-ray fluorescence analysis of water. Anal. Chem. *49*, 1326 (1977)

Received January 4, 1979

Author Index Volumes 26–85

The volume numbers are printed in italics

Albini, A., and Kisch, H.: Complexation and Activation of Diazenes and Diazo Compounds by Transition Metals. *65*, 105–145 (1976).

Altona, C., and Faber, D. H.: Empirical Force Field Calculations. A Tool in Structural Organic Chemistry. *45*, 1–38 (1974).

Anderson, D. R., see Koch, T. H.: *75*, 65–95 (1978).

Anderson, J. E.: Chair-Chair Interconversion of Six-Membered Rings. *45*, 139–167 (1974).

Anet, F. A. L.: Dynamics of Eight-Membered Rings in Cyclooctane Class. *45*, 169–220 (1974).

Ariëns, E. J., and Simonis, A.-M.: Design of Bioactive Compounds. *52*, 1–61 (1974).

Aurich, H. G., and Weiss, W.: Formation and Reactions of Aminyloxides. *59*, 65–111 (1975).

Balzani, V., Bolletta, F., Gandolfi, M. T., and Maestri, M.: Bimolecular Electron Transfer Reactions of the Excited States of Transition Metal Complexes. *75*, 1–64 (1978).

Bardos, T. J.: Antimetabolites: Molecular Design and Mode of Action. *52*, 63–98 (1974).

Barnes, D. S., see Pettit, L. D.: *28*, 85–139 (1972).

Bauder, A., see Frei, H.: *81*, 1–98 (1979).

Bastiansen, O., Kveseth, K., and Møllendal, H.: Structure of Molecules with Large Amplitude Motion as Determined from Electron-Diffraction Studies in the Gas Phase. *81*, 99–172 (1979).

Bauer, S. H., and Yokozeki, A.: The Geometric and Dynamic Structures of Fluorocarbons and Related Compounds. *53*, 71–119 (1974).

Baumgärtner, F., and Wiles, D. R.: Radiochemical Transformations and Rearrangements in Organometallic Compounds. *32*, 63–108 (1972).

Bayer, G., see Wiedemann, H. G.: *77*, 67–140 (1978).

Bernardi, F., see Epiotis, N. D.: *70*, 1–242 (1977).

Bernauer, K.: Diastereoisomerism and Diastereoselectivity in Metal Complexes. *65*, 1–35 (1976).

Bikerman, J. J.: Surface Energy of Solids. *77*, 1–66 (1978).

Boettcher, R. J., see Mislow, K.: *47*, 1–22 (1974).

Bolletta, F., see Balzani, V.: *75*, 1–64 (1978).

Brandmüller, J., and Schrötter, H. W.: Laser Raman Spectroscopy of the Solid State. *36*, 85–127 (1973).

Bremser, W.: X-Ray Photoelectron Spectroscopy. *36*, 1–37 (1973).

Breuer, H.-D., see Winnewisser, G.: *44*, 1–81 (1974).

Brewster, J. H.: On the Helicity of Variously Twisted Chains of Atoms. *47*, 29–71 (1974).

Brocas, J.: Some Formal Properties of the Kinetics of Pentacoordinate Stereoisomerizations. *32*, 43–61 (1972).

Brown, H. C.: Meerwein and Equilibrating Carbocations. *80*, 1–18 (1979).

Brunner, H.: Stereochemistry of the Reactions of Optically Active Organometallic Transition Metal Compounds. *56*, 67–90 (1975).

Buchs, A., see Delfino, A. B.: *39*, 109–137 (1973).

Bürger, H., and Eujen, R.: Low-Valent Silicon. *50*, 1–41 (1974).

Burgermeister, W., and Winkler-Oswatitsch, R.: Complexformation of Monovalent Cations with Biofunctional Ligands. *69*, 91–196 (1977).

Burns, J. M., see Koch, T. H.: *75*, 65–95 (1978).

Inorganic Chemistry Concepts

Editors:
M. Becke
C. K. Jørgensen
M. F. Lappert
S. J. Lippard
J. L. Margrave
K. Niedenzu
R. W. Parry
H. Yamatera

Volume 1
R. Reisfeld, C. K. Jørgensen

Lasers and Excited States of Rare Earths

1977. 9 figures, 26 tables. VIII, 226 pages
ISBN 3-540-08324-3

Contents:
Analogies and Differences Between Monatomic Entities and Condensed-Matter. – Rare-Earth Lasers. – Chemical Bonding and Lanthanide Spectra. – Energy Transfer. – Applications and Suggestions.

Volume 2
R. L. Carlin, A. J. van Duyneveldt

Magnetic Properties of Transition Metal Compounds

1977. 149 figures, 7 tables. XV, 264 pages
ISBN 3-540-08584-X

Contents:
Paramagnetism: The Curie Law. – Thermodynamics and Relaxation. – Paramagnetism: Zero-Field Splittings. – Dimers and Clusters. – Long-Range Order. – Short-Range Order. – Special Topics: Spin-Flop, Metamagnetism, Ferrimagnetism and Canting. – Selected Examples.

Volume 3
P. Gütlich, R. Link, A. Trautwein

Mössbauer Spectroscopy and Transition Metal Chemistry

1978. 160 figures, 1 folding plate, 19 tables.
X, 280 pages
ISBN 3-540-08671-4

Contents:
Basic Physical Concepts. – Hyperfine Interactions. – Experimental. – Mathematical Evaluation of Mössbauer Spectra. – Interpretation of Mössbauer Parameters of Iron Compounds. – Mössbauer-Active Transition Metals Other than Iron. – Some Special Applications.

Springer-Verlag
Berlin
Heidelberg
New York

H. Engelhardt

High Performance Liquid Chromatography

Chemical Laboratory Practice

Translated from the German by G. Gutnikov
1979. 73 figures, 13 tables. XII, 248 pages.
ISBN 3-540-09005-3

Contents:
Chromatographic Processes. – Fundamentals of Chromatography. – Equipment for HPLC. – Detectors. – Stationary Phases. – Adsorption Chromatography. – Partition Chromatography. – Ion-Exchange Chromatography. – Exclusion Chromatography. Gel Permeation Chromatography. – Selection of the Separation System. – Special Technique. – Purification of Solvents. – Subject Index.

This simple and non-mathematical introduction to high-performance-liquid chromatography (HPLC) emphasizes the practical aspects of achieving a successful separation. This method usually permits analyses to be carried out more rapidly than by gas chromatography and is, more-over, eminently suited for the separation of heatlabile, high-boiling, or non-volatile substances, without lengthy or tedious derivatization. In principle, all substances that are stable in solution are amenable to separation by HPLC.

HPLC equiment is described in terms of the individual components, their expected performance capabilities and suitability for certain applications.

The areas of applications of the various separation techniques (adsorption, partition, ion-exchange, exclusion) are pointed out in order to facilitate selection of the most appropriate technique by the worker for his particular problem. Considerable discussion is devoted to the parameters that are important in optimizing or improving a given separation.

The application of HPLC to actual problems in organic chemistry, pharmacological research, medicine, biochemistry and petrochemistry are illustrated by numerous relevant examples. This book is a translation of the wellknown and very successful German edition.

K. Cammann

Working with Ion-Selective Electrodes

Chemical Laboratory Practice

Translated from the German by A. H. Schroeder
1979. 65 figures. Approx. 240 pages.
ISBN 3-540-09320-6

Contents:
Fundamentals of Potentiometry. – Electrode Potential Measurements. – Ion-Selective Electrodes. – Measuring Techniques with Ion-Selective Electrodes. – Analysis Techniques Using Ion-Selective Electrodes. – Applications of Ion-Selective Electrodes. – Appendix.

The field of ion-selective electrodes has grown enormously since the publication of the first edition of this work. The Second Edition, now in English, considers new developments which have since taken place in gas sensors, enzyme electrodes and industrial applications of ion-selective electrodes.

Lucidly written and containing a helpful index, the book uses a new theoretical approach to explain the behavior of electrodes in a way comprehensible to a new-comer in the field. Various electrode types available are described as well as "do-it-yourself" electrodes, micro-electrodes, industrial flow-thru assemblies, and pollution control monitors.

The accent of the book is on simple procedures. The user therefore learns to avoid errors resulting from an unfamiliarity with electrochemical measurements of single ion activities. The bibiliography of the New Edition has been expanded considerably to include important new publications.

Already regarded as a standard among the literature in West Germany, this new English Edition now offers analytical scientists world-wide a systematic and inviting introduction to the field of ion-selective electrodes.

Springer-Verlag
Berlin Heidelberg New York